An Official Journal of
International Union of Nutritional Sciences (IUNS)
Federation of European Nutrition Societies (FENS)
European Academy of Nutritional Sciences (EANS)

Annals of Nutrition & Metabolism

Journal of Nutrition, Metabolic Diseases and Dietetics

Founded 1959 as 'Nutritio et Dieta' by E. Azerad, H. Kapp and J. Trémolières
Continued by A. Wretlind (1961–1969)
Continued by N. Zöllner (1970–1990) as 'Nutrition and Metabolism' (1970–1980), since 1980 integrating 'Annales de la Nutrition et de l'Alimentation', continued as 'Annals of Nutrition and Metabolism'
Continued by G. Wolfram (1991–1999)

KARGER

Printed in Switzerland on acid-free paper by Reinhardt Druck, Basel

Appears bimonthly: 1 volumes per year (6 issues)

Traditional uses and health beneficial effects of selected local food plants

Local Mediterranean Food Plants and Nutraceuticals

Editors
Michael Heinrich
Walter E. Müller
Claudio Galli

Nutrition, Clinical Nutrition, Dietetics, Anthropology, Botany, Micronutrients, Phytomedicine, Pharmacology, Gerontology, Cardiovascular System, Oncology

As a result of domestication – considered to be the most important cultural development of the past 13,000 years of human history - we depend today on a tiny number of domesticated plant and animal species for our food supply. Nevertheless, people continue to gather food which grows around their homes or they cultivate local varieties of food plants (as well as keep land races of domestic animals). Generally, wild varieties tend to be richer in micronutrients and bioactive secondary metabolites, which are produced in adaptation to local environmental conditions. These metabolites trigger further adaptive responses by producing 'protective', bioactive compounds which, when ingested, result in the transfer of protective effects to our organism.

The preservation of local knowledge for future generations as well as the identification of new nutraceuticals by means of characterizing plant extracts with potential health beneficial effects are the main objectives of this book. The first part therefore focuses on the ethnobotanical study of local food plants in selected regions of the Mediterranean. The biological-pharmacological effects of these plants on selected (mostly in vitro) targets of the central nervous system and the cardiovascular system make up the second part, highlighting the potential of these plants for developing novel health foods, herbal medicines or local products with beneficial effects on health.

Those interested in an integrated approach to the identification of new and health-beneficial foods will undoubtedly find this book a valuable source of information and an inspiration for new scientific approaches to this age-old topic.

KI 06650

Contents

www.karger.com/fonut

Forum of Nutrition, Vol. 59
Series Editor: Elmadfa, I. (Vienna)
ISSN 1660–0347

Local Mediterranean Food Plants and Nutraceuticals
Editors: Heinrich, M. (London); Müller, W.E. (Frankfurt); Galli, C. (Milano)
XII + 186 p., 24 fig., 23 tab., hard cover, 2006
CHF 228.– / EUR 163.– / USD 207.25
Prices subject to change
EUR price for Germany, USD price for USA only
ISBN-10: 3–8055–8124–6
ISBN-13: 978–3–8055–8124–0

KARGER

Order Form

Please send: ______ copy/ies

Postage and handling free with prepayment

Payment:
☐ Check enclosed ☐ Please bill me

Please charge to my credit card
☐ American Express ☐ Diners ☐ Eurocard
☐ MasterCard ☐ Visa

Card No.: ______________________

Exp. date: ______________________

Orders may be placed with any bookshop, subscription agency, directly with the publisher or through a Karger distributor.

Fax: +41 61 306 12 34

S. Karger AG, P.O. Box, CH–4009 Basel (Switzerland)
E-Mail orders@karger.ch, **www.karger.com**

Name/Address: ______________________

Date: ______________________

Signature: ______________________

General Information

Annals of
Nutrition & Metabolism

ISSN Print Edition: 0250–6807
ISSN Online Edition: 1421–9697

Journal Homepage: www.karger.com/anm

Publication Data: 'Annals of Nutrition and Metabolism' is published 6 times a year. Volume 51 with 6 issues appears in 2007.

Subscription Rates: Subscriptions run for a full calendar year. Prices are given per year.

Personal subscription:

Print or Online	Print+Online combined
CHF 571.20	CHF 616.20
EUR 423.–	EUR 456.–
USD 519.30	USD 560.30

postage and handling (added to print and print+online)
CHF 34.50 Europe, CHF 51.30 Overseas
EUR 24.90
USD 47.10

Institutional subscription:

Print or Online	Print+Online combined
CHF 1904.–	CHF 2094.–
EUR 1410.–	EUR 1551.–
USD 1731.00	USD 1904.00

postage and handling (added to print and print+online)
CHF 43.20 Europe, CHF 64.20 Overseas
EUR 31.20
USD 58.80

Airmail surcharge: CHF 43.50 / USD 39.30

Discount subscription prices:
Please enquire about reduced rates for members of affiliated societies.

Back Volumes and Single Issues: Information on availability and prices of single print issues and print or electronic back volumes can be obtained from Customer Service at *service@karger.ch.*

Bibliographic Indices: This journal is regularly listed in bibliographic services, including *Current Contents®* and PubMed/MEDLINE.

Photocopying: This journal has been registered with the Copyright Clearance Center (CCC), as indicated by the code appearing on the first page of each article. For readers in the US, this code signals consent for copying of articles for personal or internal use, or for the personal or internal use of specific clients, provided that the stated fee is paid per copy directly to
Copyright Clearance Center Inc.
222 Rosewood Drive
Danvers, MA 01923 (USA)
A copy of the first page of the article must accompany payment. Consent does not extend to copying for general distribution, for promotion, for creating new works, or for resale. In these cases, specific written permission must be obtained from the copyright owner,
S. Karger AG, P.O. Box
CH–4009 Basel (Switzerland).

Subscription Orders:
Orders can be placed at agencies, bookstores, directly with the Publisher or further Karger offices or representatives:

S. Karger AG
Medical and Scientific Publishers
P.O. Box
CH–4009 Basel
Switzerland
(for courier services only:
Allschwilerstrasse 10
CH–4055 Basel)
Tel. +41 61 306 11 11
Fax +41 61 306 12 34
E-Mail karger@karger.ch
www.karger.com

France:
Librairie Medi-Sciences Sarl
36, bd de Latour-Maubourg
75007 Paris
France
Tél. +33 (0) 1 45 51 42 58
Fax +33 (0) 1 45 56 07 80
E-Mail librairie@medi-sciences.fr
www.medi-sciences.fr

Germany:
S. Karger GmbH
Postfach
79095 Freiburg
Deutschland
(Hausadresse: Lörracher Strasse 16A
79115 Freiburg)
Tel. +49 761 45 20 70
Fax +49 761 45 20 714
E-Post information@karger.de
www.karger.de

India, Bangladesh, Sri Lanka:
Panther Publishers Private Ltd.
33, First Main
Koramangala First Block
Bangalore 560 034
India
Tel. +91 80 25505 836
Tel. +91 80 25505 837
Fax +91 80 25505 981
E-Mail panther_publishers@vsnl.com
www.pantherpublishers.com

Japan:
Karger Japan, Inc.
Yushima S Bld. 3F
4-2-3, Yushima, Bunkyo-ku
Tokyo 113-0034
Japan
Tel. +81 3 3815 1800
Fax +81 3 3815 1802
E-Mail publisher@karger.jp

People's Republic of China:
Creative Consulting (Shanghai) Co., Ltd.
1 Fuxing Zhong Lu
Suite 1208
Shanghai 200021
China
Tel. (86 21) 63 91 92 90
Fax (86 21) 63 91 92 95
E-Mail karger_china@yahoo.com.cn

South America and Central America:
Cranbury International LLC
7 Clarendon Ave., Suite 2
Montpelier, VT 05602
USA
Tel. +1 802 223 6565
Fax +1 802 223 6824
E-Mail
eatkin@cranburyinternational.com
www.cranburyinternational.com

United Kingdom, Ireland:
S. Karger AG
c/o London Liaison Office
4 Rickett Street
London SW6 1RU
United Kingdom
Tel. +44 (0) 20 7386 0500
Fax +44 (0) 20 7610 3337
E-Mail uk@karger.ch

USA:
S. Karger Publishers, Inc.
26 West Avon Road
P.O. Box 529
Unionville, CT 06085
USA
Toll free: +1 800 828 5479
Tel. +1 860 675-7834
Fax +1 860 675-7302
E-Mail karger@snet.net

Change of Address:
Both old and new address should be sent to the subscription source.

KARGER

Fax +41 61 306 12 34
E-Mail karger@karger.ch
www.karger.com

The Journal Home Page is available at:
www.karger.com/anm

Challenging the cholesterol hypothesis

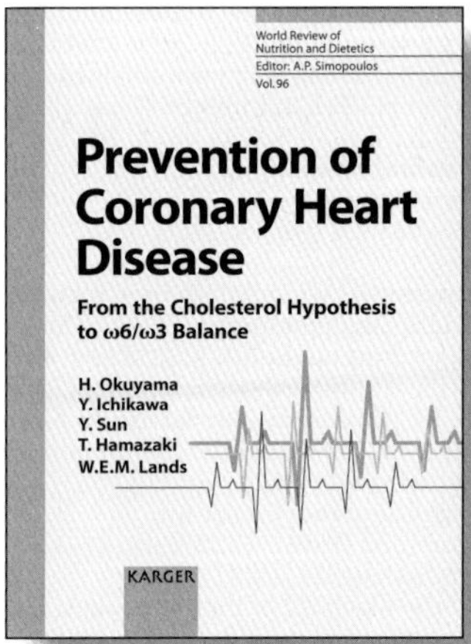

Prevention of Coronary Heart Disease

From the Cholesterol Hypothesis to ω6/ω3 Balance

Harumi Okuyama, Yuko Ichikawa, Yueji Sun, Tomohito Hamazaki, William E.M. Lands

Nutrition; Cardiovascular System; Preventive Medicine; Clinical Nutrition, Dietetics, Angiology, Arteriosclerosis, Biochemsitry, Epidemiology

Main Contents

- The Cholesterol Hypothesis – Its Basis and Its Faults
- Association of High Total Cholesterol with Coronary Heart Disease Mortality Differs among Subject Populations – Familial Hypercholesterolemia as a Key Concept
- Cancer and All-Cause Mortalities Are Lower in the Higher Total Cholesterol Groups among General Populations
- Pleiotropic Effects of Statins in the Prevention of Coronary Heart Disease – Potential Side Effects
- Objective Measures of the Pathology of Coronary Heart Disease
- ω3 Fatty Acids Effectively Prevent Coronary Heart Disease and Other Late-Onset Diseases – The Excessive Linoleic Acid Syndrome
- Why Isn't the Causal Relationship between Linoleic Acid and Mortalities from Coronary Heart Disease and Stroke Revealed by Clinical Studies?
- Mechanisms by Which Dietary Fats Affect Coronary Heart Disease Mortality
- Cancers Common in the USA Are Stimulated by ω6 Fatty Acids and Large Amounts of Animal Fats, but Suppressed by ω3 Fatty Acids and Cholesterol
- New Directions of Lipid Nutrition for the Primary and Secondary Prevention of Coronary Heart Disease and Other Late-Onset Diseases

This publication is organized in an exceptional way: Each chapter introduces several completed clinical trials and provides the original conclusions and discussions of the results. The authors then contribute their own comments and interpretations of the findings, challenging the prevailing belief that serum cholesterol is a mediator of disease which is increased by eating saturated fats and decreased by eating polyunsaturated fats. They argue that upon closer scrutiny, the diet recommendations based on the cholesterol hypothesis are essentially ineffective in reducing serum cholesterol levels in the long run. Instead, it is proposed that traditional cholesterol biomarkers are of different significance in short- and long-term interventions due to the feedback control mechanisms in the body. Even more important, the association of high serum cholesterol values with high coronary heart disease mortality is not consistent when different populations are compared: This mortality rate may simply reflect the incidence and severity of familial hypercholesterolemia cases. This agrees with the observation that higher serum cholesterol values associate with lower cancer and all-cause mortalities in populations with a low relative proportion of this disorder. Thus, there seems to be no benefit of limiting dietary cholesterol intake or lowering serum cholesterol values below a certain limit. Moreover, evidence has been found that the health risk results from high intakes of calories, aggravated by an unbalanced intake of ω6/ω3 polyunsaturated fats.
Based on the reviewed data, new directions of lipid nutrition are recommended for the primary and secondary prevention of coronary heart disease, cancer and all-cause deaths, which will likely revolutionize current dietary practice.

KI 06667

www.karger.com/wrund

World Review of Nutrition and Dietetics, Vol. 96
Series Editor: Simopoulos, A.P. (Washington, D.C.)
ISSN 0084-2230

Prevention of Coronary Heart Disease
From the Cholesterol Hypothesis to ω6/ω3 Balance
Okuyama, H.; Ichikawa, Y. (Nagoya); Sun, Y. (Dalian); Hamazaki, T. (Toyama); Lands, W.E.M. (College Park, Md.)
X + 168 p., 92 fig., 89 in color, 23 tab., hard cover, 2007
CHF 228.– / EUR 163.– / USD 207.25
Prices subject to change
EUR price for Germany, USD price for USA only
ISBN-10: 3–8055–8179–3
ISBN-13: 978–3–8055–8179–0

KARGER

Order Form

Please send: _____ copy/ies

Postage and handling free with prepayment

Payment:
☐ Check enclosed ☐ Please bill me

Please charge to my credit card
☐ American Express ☐ Diners ☐ Eurocard
☐ MasterCard ☐ Visa

Card No.: ____________

Exp. date: ____________

Orders may be placed with any bookshop, subscription agency, directly with the publisher or through a Karger distributor.

Fax: +41 61 306 12 34

S. Karger AG, P.O. Box, CH–4009 Basel (Switzerland)
E-Mail orders@karger.ch, **www.karger.com**

Name/Address: ____________

Date: ____________

Signature: ____________

Annals of
Nutrition&
Metabolism

Food-Based Dietary Guidelines – Practical Aspects of Implementation

Guest Editors
P. Walter, Basel
I. Elmadfa, Wien

19 figures, 10 in color, and 17 tables, 2007

KARGER
Basel · Freiburg · Paris · London · New York ·
Bangalore · Bangkok · Singapore · Tokyo · Sydney

This edition was made possible by the generous support of ISFE.
An AMC foundation

An AMC foundation

S. Karger
Medical and Scientific Publishers
Basel • Freiburg • Paris • London
New York • Bangalore • Bangkok
Singapore • Tokyo • Sydney

Disclaimer
The statements, options and data contained in this publication are solely those of the individual authors and contributors and not of the publisher and the editor(s). The appearance of advertisements in the journal is not a warranty, endorsement, or approval of the products or services advertised or of their effectiveness, quality or safety. The publisher and the editor(s) disclaim responsibility for any injury to persons or property resulting from any ideas, methods, instructions or products referred to in the content or advertisements.

Drug Dosage
The authors and the publisher have exerted every effort to ensure that drug selection and dosage set forth in this text are in accord with current recommendations and practice at the time of publication. However, in view of ongoing research, changes in government regulations, and the constant flow of information relating to drug therapy and drug reactions, the reader is urged to check the package insert for each drug for any change in indications and dosage and for added warnings and precautions. This is particularly important when the recommended agent is a new and/or infrequently employed drug.

Printed in Switzerland on acid-free paper by
Reinhardt Druck, Basel
ISBN 978–3–8055–8328–2

KARGER

Fax +41 61 306 12 34
E-Mail karger@karger.ch
www.karger.com

Vol. 51, Suppl. 2, 2007

Annals of
Nutrition&
Metabolism

Contents

KARGER

Fax +41 61 306 12 34
E-Mail karger@karger.ch
www.karger.com

Access to full text and tables of contents, including tentative ones for forthcoming issues:
www.karger.com/anm_issues

Annals of
Nutrition&
Metabolism

Ann Nutr Metab 2007;51(suppl 2):1
DOI: 10.1159/000103559

Published online: July 30, 2007

Food-Based Dietary Guidelines – Practical Aspects of Implementation

16th International ISFE Symposium, Vienna, June 28, 2006

When we look back about 60 years, nutrient deficiencies were a major problem in Europe. Towards the end of the last century, deficiencies were much less frequent, but the upcoming abundance of food resulted in an increase in cardiovascular disease and stroke. At the present time, too many people all over the world are eating too much and are physically inactive. As a result, obesity is increasing dramatically and is now recognized as one of the most serious problems in the industrialized as well as developing world. In both cases, neither nutrition nor lifestyle was and is appropriate. Actually, it is now well established what we should eat to stay healthy, but people do not listen or do not believe in the sometimes contradictory recommendations that are issued by various organizations and by the media. Therefore, the responsible scientific organizations and the governments have to improve their communication on how and what to eat. In 1996, FAO/WHO published that the dietary recommendations have to be practical, to be comprehensible, and to be culturally acceptable. The key concept for communication needs to be reconsidered. If possible, such food-based guidelines should be visualized in a simple way. In this symposium, some of these questions are addressed. Which types of concepts and ways of communication have been developed? Models of balanced diets have been proposed and visualized such as the Nutrition Circle, Food Pyramid or in the form of campaigns like '5-a-day'. The goal of this symposium is to look at the practical aspects of implementation and communication of these nutritional messages.

This symposium was organized by The International Foundation for the Promotion of Nutrition Research and Nutrition Education (ISFE) which is supported by AMC International Alfa Metalcraft Corporation AG, a company that develops and sells high-quality products for gentle cooking without added water and fat. The organizers thank AMC for supporting this symposium.

Paul Walter, Basel
Ibrahim Elmadfa, Vienna

KARGER

Fax +41 61 306 12 34
E-Mail karger@karger.ch
www.karger.com

0250–6807/07/0518–0001$23.50/0

Accessible online at:
www.karger.com/anm

Annals of
Nutrition&
Metabolism

Ann Nutr Metab 2007;51(suppl 2):2–7
DOI: 10.1159/000103560

Published online: July 30, 2007

Global Patterns and Country Experiences with the Formulation and Implementation of Food-Based Dietary Guidelines

Janice Albert

Nutrition and Consumer Protection Division, Food and Agriculture Organization, Rome, Italy

Key Words

Food-based dietary guidelines, global · Nutrition education · National dietary guidelines · Dietary guidelines · FAO, developing country diets

Abstract

Nutrition information and advice for the public must be credible from a scientific perspective and easy to understand and follow. The Food and Agriculture Organization of the United Nations (FAO) has promoted the preparation and use of food-based dietary guidelines (FBDG) in Asia, the Near East, Latin America and the Caribbean and Africa. The paper is based on a survey of 28 developing countries and case studies of South Africa, India, Chile and Guatemala, which were conducted by FAO in 2004. All FBDG stress the importance of dietary variety and balance, promote fruit and vegetable consumption, discourage excessive salt, sugar and alcohol consumption, and promote physical activity. In other respects, FBDG vary according to the specific health, behavior, culture and economic conditions within a country. The steps used in developing and implementing FBDG are discussed, with country examples to illustrate achievements, mistakes and obstacles. Main areas for improvements are: (1) FBDG are mainly developed and used within the health sector, involvement of the agriculture and education sectors is needed; (2) few counties communicate to widely reach the public; (3) many messages do not motivate changes in behavior, and (4) the application of FBDG in setting policies not sufficient received attention.

The views expressed in this publication are those of the author and do not necessarily reflect the views of the Food and Agriculture Organization of the United Nations.

Background/Aims

There is keen public interest in nutrition in many countries and it is important that the information and advice that the public receives is credible from a scientific perspective and easy to understand and follow. Food-based dietary guidelines (FBDG) aim to provide this type of information to the general public. FBDG use common language and are intended for use by the average person; they should not be confused with recommended nutrient intakes which provide quantitative recommendations for use by specialists for ensuring the health of populations [1]. While recommended nutrient intakes are standards which apply worldwide, FBDG are specifically designed for each country and advice can differ according to the needs of the particular population. FBDG are also a policy tool, which can be used to coordinate nutrition programs and to evaluate the adequacy of the food supply in meeting the nutritional needs of the population [2].

The Food and Agriculture Organization of the United Nations (FAO) and the World Health Organization

KARGER

Fax +41 61 306 12 34
E-Mail karger@karger.ch
www.karger.com

Janice Albert
Nutrition and Consumer Protection Division
Food and Agriculture Organization
Viale delle Terme di Caracalla, IT–00153 Rome (Italy)
Tel. +39 06 570 53552, Fax +39 06 570 54593, E-Mail Janice.Albert@fao.org

(WHO) have promoted the notion that governments should provide nutritional advice to the public which is relevant for different age groups and lifestyles, and culturally appropriate to the country's population [3]. In 1995, FAO and WHO sponsored an Expert Consultation on the Preparation and Use of Food-Based Dietary Guidelines [4]. The technical report of this meeting provided the rationale for FBDG and an overview of the steps involved in producing dietary guidelines. After the meeting, FAO sponsored 11 workshops in which more than 95 countries participated. WHO promoted the concept of FBDG, primarily through its regional offices. The two UN agencies have trained nutritionists and facilitated the development of FBDG in Asia, the Near East, Latin America and the Caribbean and Africa.

This aim of this paper is to describe current global patterns in the contents of FBDG and to analyze countries' experiences with the formulation and implementation of FBDG.

Method

Recognizing the large gaps in available information about the development and uses of FBDG in the developing world, FAO initiated a pilot study of FBDG in August 2004. Our research included a survey and four of case studies.

Survey

A questionnaire containing broad, open-ended questions was prepared in English and Spanish and sent by e-mail to 42 countries. These countries were chosen because they had sent government staff to participate in FAO workshops on FBDG. Twenty-eight countries responded to the survey (response rate of 67%).

Government representatives were asked if they had developed FBDG and food guides, what messages were contained in the FBDG, which organizations and individuals were involved in preparation of the FBDG, how the FBDG were being used in the country, if there were problems in using the FBDG, and whether the FBDG had been evaluated or there were plans to evaluate their use.

Case Studies

Five experts in nutrition education from India, South Africa, Chile and Guatemala were commissioned to provide case studies of their countries' experiences in producing and disseminating FBDG. These countries were chosen because they have highly diverse populations and a range of very challenging nutrition education needs. The experts used a range of methods including face-to-face interviews, telephone interviews, surveys and review of documentation. Since four of the experts had been directly involved in the development of their countries' FBDG, they were also able to draw upon personal in-depth knowledge of the processes followed.

Results

The information provided below is based on the research study about FBDG, as well as FAO experiences in assisting countries in developing their FBDG and food guides.

Global Patterns

Throughout the world, there are several characteristics commonly found in national dietary guidelines. First, the importance of variety and balance are always stressed in FBDG. Second, it is very common for FBDG to promote increased fruit and vegetable consumption. Third, FBDG often include advice to discourage excessive salt, sugar and alcohol consumption. Fourth, with increased concern about sedentary lifestyles, physical activity messages are becoming more common.

Close examination of dietary guidelines demonstrates that there are significant differences among FBDG. This is not surprising since FBDG are determined by the specific health, behavior, culture and economic conditions within a country. For example, advice about dietary fats and carbohydrates may vary greatly, according to the nutritional problems in a country.

Providing one set of guidelines for an entire nation is challenging and sometimes controversial. Recognizing the different nutritional needs during the life cycle, a few countries have created separate dietary guidelines for different age groups (e.g. Brazil); while others have incorporated advice for specific target groups in their general guidelines (e.g. Bangladesh and Nigeria). In very large countries with extremely diverse populations, the notion of separate guidelines for different segments of the population has arisen at times. Interestingly, all of the countries in our study, no matter how large and diverse, had created FBDG which were intended for everyone. Whether these general FBDG have been effective among different population groups is a matter for future research. The presentation of the messages varies as well, some countries have very short simple messages while others give fairly detailed advice. To view the diversity of FBDG and food guides from around the world, see the FAO website: http://www.fao.org/ag/agn/nutrition/education_guidelines_en.stm.

Process of Developing FBDG

FAO and WHO recommend a series of steps to develop FBDG. Each step is important for the efficient and effective development of dietary guidelines (fig. 1). The

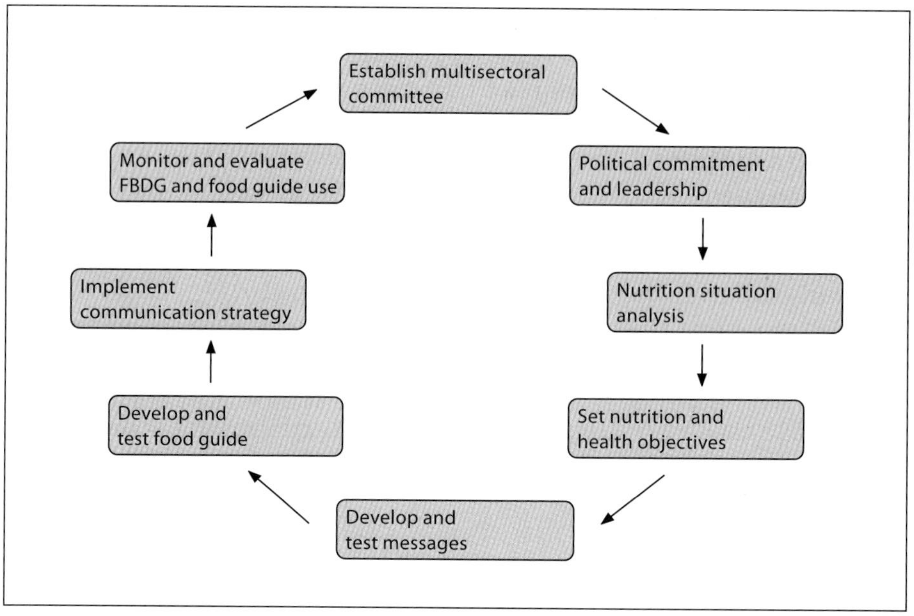

Fig. 1. Recommended steps for developing and implementing FBDG.

process is dynamic: dietary guidelines and food guides should not considered to be permanent. They are tools which should be reassessed periodically to ensure that they are always up to date and relevant to the public's needs. The process applies methods that can seem overwhelming and complicated. However, experiences from many types of countries show that the challenges of developing FBDG can be met even in countries which lack extensive resources.

Multisectoral Committee

A first step in developing FBDG is the formation of a national committee which includes representatives of the leading ministries (e.g. health, agriculture, education, social welfare) and others who can assist in developing and promoting FBDG such as scientific societies, food industry associations and consumer organizations. While including a range of institutions is essential, experience shows that FBDG development is most successful when there is a leading individual who is able to dedicate time to advocating the concept of FBDG and coordinating the process.

Example – Guatemala: When Guatemala began work on its dietary guidelines in 1995, they formed a commission with support from the Department of Nutrition within the Ministry of Public Health and Social Assistance. Schools of education, a pediatric hospital, the Department of Maternal and Child Health, the Ministry of Education, the Ministry of Agriculture, Ranching and Food as well as the PAHO Institute of Nutrition for Central America and Panama were members of the commission. The multisectoral team met over a 2-year period to review evidence about the nutrition situation and to decide on the main objectives of the guidelines. They approved the messages and food guide and committed their institutions to promoting the dietary guidelines [5].

Political Support and Commitment

Without political support and coordination, bureaucracies may not incorporate the FBDG into their programs or devote resources to developing and promoting the FBDG. In a number of countries, the Parliament approves the FBDG and the Prime Minister or President announces the guidelines.

Example – South Africa: South Africa's experience illustrates the importance of involving all the relevant government agencies and stakeholders at the earliest stages in developing FBDG. In 1997, the Nutrition Society of South Africa began developing FBDG. A committee included universities, the food industry, health profession associations, a UN agency and the Directorate of Nutrition within the Ministry of Health. Participation in the committee was voluntary and participation from some representatives was low [6]. Before the FBDG were approved by the Parliament, a public controversy arose over some aspects of the recommendations regarding meat and carbohydrates. This delayed the approval of the FBDG until 2003. Better participation and involvement of all the relevant government units and stakeholders earlier in the process might have prevented this delay.

Analysis of Nutrition Situation and Setting of Objectives

The next steps are technical: the analysis of the nutrition situation and identification of objectives is carried out by nutritionists and other specialists. Using data such as health and census statistics, household expenditure and food consumption surveys, food balance sheets, and the scientific literature, the technical analysis leads to identification of the nutrition issues which are significant for public health. Most countries do not have good, comprehensive data for formulating their FBDG and the lack of data can cause long delays in developing FBDG.

Example – Namibia: Namibia's experiences with developing their FBDG illustrate how a country coped with the lack of information. In 1998, FAO provided technical assistance to Namibia for the formulation of their guidelines. The Health Ministry did not have information about food consumption. To carry out their analysis, they relied on data on household expenditures and the observations of medical personnel to obtain information about the most important health problems and priorities [7].

The positive lesson from Namibia is that the country found a way to cope with the limitations in their resources and followed the recommended steps to the extent possible. They did not allow an imperfect situation to stop their FBDG development.

Testing of Messages and Food Guides

The general public has difficulty understanding the words and concepts which are used by nutritionists. Even after the guidelines are developed, the words used in public information campaigns may have to be modified to be understood by different audiences [8].

Some concepts which are commonly used by nutritionists are not easily conveyed to all the segments of the public. For example, dietary variety and portion size are basic concepts in nutrition but they are often misunderstood. Consumers may not understand the amounts conveyed by words such as 'sparingly', 'increase', 'moderate', and visual symbols and common measurements can assist them. The testing of messages and food guides is a crucial step in the formulation of FBDG. Unfortunately, countries may omit this step due to lack of awareness of its importance or lack of resources to carry out this type of research. Regardless of the amount of expertise and experience existing within a team, pre-testing the messages and food guides always brings valuable new information.

Example – Thailand: Before Thailand published its FBDG and Nutrition Flag in 1998, they tested the food guide and messages. The public was asked about the Nutrition Flag to assess their understanding of portion size and quantities. Thais developed the 'rice-serving spoon' as the household unit for measuring foods. Using this flat broad spoon, the research team obtained realistic estimates of portions eaten and a way to explain quantities to the public. Hundreds of people were interviewed in department stores, food markets, factories and universities. Since travel to rural areas was difficult, the researchers overcame this problem by interviewing commuters from rural areas at bus stations [9].

Communication Strategy and Dissemination of Information

A major obstacle in the implementation of FBDG is that nearly all countries lack sufficient resources to develop and disseminate FBDG-related information. Without a communication strategy, the FBDG materials may not reach a wide audience and they may have a very limited impact. Education and communication expertise, as well as nutrition expertise, is needed for the successful formulation and implementation of FBDG. FBDG must be widely distributed and community workers must be trained to use the materials.

Example – India: When India published its dietary guidelines in 1998, the materials were translated into 6 Indian languages to reach the diverse population and there was newspaper, radio and television coverage in the various regions of the country. India used its extensive health, agriculture and education networks for disseminating information. Professionals and community workers received training about the FBDG. Non-governmental organizations and professional societies also promoted the guidelines. Special events such as 'national nutrition week' and 'agriculture field days' were used to promote the guidelines [10]. A key lesson is that countries should use the well-established channels of communication and existing training activities to promote the FBDG.

Monitoring and Evaluation

The process of disseminating FBDG and food guides should be assessed to ensure that the messages are reaching their audiences. Without evaluation, it is difficult to demonstrate that the messages and food guide have an impact on consumer awareness and to know whether knowledge, attitudes and behaviors are changing. The evaluations should take into consideration the environment and potential for change within realistic time frames. Expertise is needed to conduct an evaluation of FBDG. Many countries in our survey had not conducted evaluations and did not know how to carry out this activity.

Example – Chile: Chile is one of the few countries which has evaluated its dietary guidelines and food guide. The evaluation included different types of institutional settings and population groups. A study in a clinic found that patients with diet-related chronic diseases received messages to eat more fruits and vegetables but had not heard about reducing animal fats. A study in schools found that children had seen the pyramid but did not understand the portions. A third study surveyed consumers over the Internet. Initially, only 30% of the consumers knew about the guidelines and 60% knew about the food pyramid. After the same consumers received information directly, the awareness and willingness to change diets rose to 80% [11]. The information obtained from the evaluation was used to revise the FBDG in 2005.

Limitations in Current Knowledge

Although many countries have formulated and implemented FBDG, our current state of knowledge about these tools is very limited and there is a need for more research on dietary guidelines. Unfortunately, much of the information about the formulation and implementation of dietary advice is unavailable to the scientific community. During our pilot study, our literature search of major data bases for nutrition, health and agricultural publications identified only 50 articles which specifically addressed the topic of dietary guidelines. Few nutritionists have published information about their country's dietary guidelines in the international scientific journals. Often the papers which exist explain the messages but do not discuss how guidelines were formulated and are being used. While governments publish their guidelines, they are difficult to find and most information is published in languages which are not widely understood internationally.

Future Work

In our review of experiences with developing and implementing FBDG in different countries, we have found a number of problems. There are four main issues which could lead to improvements in the formulation and implementation of FBDG. First, our survey found that FBDG are mainly developed and used within the health sector. There is a need for greater involvement of the agriculture and education sectors. Second, few counties have a communication strategy or make use of all the ways to reach the public. Many messages do not consider the reasons people choose foods and ways to motivate changes in behavior. Fourth, FBDG are intended to assist policymakers, as well as being educational tools. However, the application of FBDG in setting policies appears to be an area which has not received attention by researchers or practitioners.

The process of developing and implementing FBDG can seem to be a daunting exercise; however, countries have found ways to overcome many obstacles. FBDG are evolving and nutritionists can benefit from sharing experiences with other countries. To make information more accessible to the scientific community and practitioners, FAO is developing a website which will provide information about FBDG and food guides around the world. In addition, we will continue to provide scientific advice and technical assistance to countries in their efforts to provide nutrition information to the public.

Acknowledgements

The author expresses appreciation to the government participants in the survey for taking time to share their experiences with FAO. In addition, the following individuals contributed to the research: Veronika Molina, Sonia Olivares, Penny Love, Anne Behr and Lalita Bhattacharjee produced case studies; Kraisid Tontisirin, William Clay, Ellen Muehlhoff, Peter Glasauer, Cecilio Morón, Dirk Schulz and Biplab Nandi contributed expertise and contacts, and Joanna Lyons, Jayne Beaney, Jenny E. Martinez Nocito and Gretchen Dueñas assisted in the research. The International Life Sciences Institute contributed financial support for the pilot research.

References

1 Taylor CL, Albert J, Weisell R, Nishida C: International dietary standards: FAO and WHO; in Bowman B, Russell R (eds): Present Knowledge in Nutrition, ed 9. Washington, International Life Sciences Institute, vol II, pp 876–887.
2 Schneeman BO: Preparation and use of food-based dietary guidelines: lessons from Thailand and the Philippines. Food Nutr Agric 2001;28:55–64.
3 FAO: International Conference on Nutrition, Plan of Action for Nutrition. Rome, FAO/WHO, 1992.
4 FAO/WHO: Preparation and Use of Food-Based Dietary Guidelines. Report of a Joint FAO/WHO Consultation. Geneva, WHO, 1998.
5 Molina V: Estudio de Caso: Proceso de Elaboracion e Implementacion de las Guias Alimentarias de Guatemala. Guatemala, INCAP/OPS, Oct 2004 (unpubl study).
6 Love P, Behr A: The South African Dietary Guidelines: Country Report. FAO, Oct 2004 (unpubl study).
7 Glasauer P: FAO Technical Backstopping Officer to the Joint Initiative of FAO, WHO, UNICEF and the Namibian Ministry of Agriculture, Water and Rural Development, Ministry of Basic Education and Culture, Ministry of Health and Social Services and University of Namibia 1998–2000 (pers commun).
8 Clay W: Preparation and use of food-based dietary guidelines. Food Nutr Agric 1997;19:42–47.
9 Dueñas G: Pretesting methods for dietary guidelines and food guide models: experiences of South Africa and Thailand. FAO, Aug 2005 (unpubl study).
10 Bhattacharjee L: Food-based dietary guidelines: case study from India. FAO, Oct 2004 (unpubl study).
11 Olivares S: Guías Alimentarias Basadas en Alimentos: La Experiencia de Chile, FAO, Oct 2004 (unpubl study).

Annals of
Nutrition&
Metabolism

Ann Nutr Metab 2007;51(suppl 2):8–14
DOI: 10.1159/000103561

Published online: July 30, 2007

Food-Based Dietary Guidelines in Austria

I. Elmadfa H. Freisling

Institute of Nutritional Sciences, University of Vienna, Vienna, Austria

Key Words
Food-based dietary guidelines · Visualizations · Nutrition policy · Austria

Abstract
The translation of nutrient-based dietary guidelines (NBDG) into food-based dietary guidelines (FBDG) is part of a nutrition policy at the national level to counteract the increasing prevalence of chronic diseases that are related to diet and physical activity. In Austria, first steps have already been taken and several promising FBDG have been introduced to the consumer. These FBDG are basically consistent as they use the same concept, a nutritive approach which is based on the 'DACH Reference Values for Nutrient Intake'. Minor differences arise in the translation of this NBDG into food servings. The 'Nutrition Circle' and pyramids are the common forms used to visualize FBDG for the Austrian population. The 'Three-Dimensional Food Pyramid', which was introduced in 2005 by the German Nutrition Society and has been adopted by the Austrian Nutrition Society, combines quantitative and qualitative aspects. A not less promising approach has been introduced by the Austrian Nutrition Society – the 'Module System' as a visual aid to facilitate the promotion and dissemination of FBDG. Together with the Austrian Nutrition Reports, FBDG are important measures to improve the nutritional situation in Austria. Upcoming results of ongoing nutrition monitoring will evaluate the success of the current measures taken in Austrian nutrition policy.

Introduction

Chronic non-communicable diseases (NCDs), including heart disease, stroke, cancer, diabetes mellitus, and obesity, caused 86% of deaths and 77% of the disease burden in the WHO European Region, yet they are largely preventable. The main modifiable risk factors include unhealthy diets and physical inactivity [1].

To reduce the burden of chronic diseases that are related to diet and physical activity, adequate nutrition policies have to be developed and maintained. The translation of nutrient-based dietary guidelines (NBDG) into food-based dietary guidelines (FBDG) is part of a nutrition policy at the national level as recommended by the World Health Organization (WHO). NBDG are directed to health professionals; for the consumer such guidelines are impractical and can hardly be translated into everyday eating practices. FBDG have proved to be an appropriate health policy measure for the improvement of nutrition. Together with specified population-based dietary goals they build the basis for an efficient nutrition information/communication [1].

Nutrition monitoring systems are essential to describe the current nutritional situation in a country, to give recommendations for an improvement of unfavorable eating habits and to track progress towards recommended dietary population goals.

The Austrian Nutrition Report 2003 [2] described the following inadequacies in current dietary habits at the population level:

- High energy intake not adjusted to the actual low energy expenditure
- High fat intake

KARGER

Fax +41 61 306 12 34
E-Mail karger@karger.ch
www.karger.com

Accessible online at:
www.karger.com/anm

Ibrahim Elmadfa
Institute of Nutritional Sciences, University of Vienna
Althanstrasse 14, AT–1090 Vienna (Austria)
Tel. +43 1 4277 54912, Fax +43 1 4277 9549
E-Mail ibrahim.elmadfa@univie.ac.at

- High intake of saturated fatty acids (of animal origin) and low intake of vegetable oils as well as the unfavorable ratio of n–6:n–3 unsaturated fatty acids
- Low intake of dietary fibers
- Low intake of folate, iodine, calcium and vitamin D.

To correct these inadequacies, several options have been proposed in the Austrian Nutrition Report, including an enhanced formulation and implementation of FBDG.

The present analysis focuses on the current situation of FBDG introduced to the Austrian population.

Table 1. Healthy eating guidelines for Austrians

1. Enjoy a variety of foods
2. Eat more carbohydrate-rich food and side dishes – eat less fried foods
3. Eat plenty of fruit and vegetables – take 5 a day
4. Eat less fat and fat-rich foods
5. Drink enough but choose your drinks reasonably
6. Enjoy your foods
7. Stay active and fit with physical activity

Healthy Eating Guidelines for Austrians

A specific chapter in the Austrian Nutrition Report 2003 [2] addressed FBDG in general and suggested guidelines for healthy eating in Austria. These guidelines are based on the results of the latest assessment of the nutritional situation in Austria. There is still need to improve the dietary behavior of the Austrian population, particularly in respect to the prevention of NCDs. In comparison to the results of the preceding Austrian Nutrition Report 1998 [3], critical nutrients and population groups at risk for inadequate nutrient intake remained mostly unchanged. Campaigns, like the '5-a-day' program to increase fruit and vegetable consumption, were able to ameliorate the dietary behavior but not in the long run. To contribute to a more lasting improvement of the diet in Austria, specifically emphasizing cultural aspects, the following guidelines have been proposed and are summarized in table 1.

1. Enjoy a variety of foods

The availability of a variety of high-quality foods over the whole year is nowadays beyond comparison to former times. Regardless of reports on occasional food scandals in media, the quality and safety of the available foods is getting better and better and with some nutritional knowledge it should be easy to choose ones most suitable foods. A diverse and balanced diet is crucial to a healthy diet. There is no single healthy, unhealthy or forbidden food; it is the balance and combination that make up a wholesome nutrition.

2. Eat more carbohydrate-rich food and side dishes – eat less fried foods

The bad reputation of the traditional Austrian diet with its abundance in sweet dishes made from white flour is somewhat iniquitous, although there are a few improvements to be made:

- Use less sugar when cooking or baking; instead prefer dried or fresh fruits as sweeteners
- Use wholemeal instead of white flour and prefer products prepared with wholemeal flour
- Fried food must not be completely avoided, but should only be eaten occasionally for example as special festive meals.

3. Eat plenty of fruit and vegetables – take 5 a day

Fruit and vegetables are crucial for a healthy diet. Many epidemiological surveys suggest that an increased intake of fruit and vegetables affords protection from a number of diseases. The variety of fruit or vegetable has less importance than good taste so that the food is enjoyed and therefore eaten. Fruit should however be fresh and vegetables not be cooked extensively. In this regard, fruit and vegetable juices are a good complement. Fruit and vegetables should ideally be part of every meal and also be chosen as snacks.

In fact, one cannot go wrong with this food class, as fruit and vegetables are always a good choice, supplying so many important nutrients, they should account for the major part of the daily diet.

4. Eat less fat and fat-rich foods

Fat, in its pure visible form (i.e. butter, margarine and oil), as well as the hidden one that is contained in many foods like sausage and cheese for instance, but also mayonnaise, sauces and dressings as well as many sweets, supplies a substantial part of our daily energy intake that is mostly too high leading to overweight in many Austrians. Moreover, fat intake can also contribute to high cholesterol levels and subsequently favor the occurrence of coronary heart disease, so that reducing the fat content of the diet is among the most important measures to prevent these risk factors and diseases.

Therefore, fat should be used sparingly in cooking and it is advisable to rely on vegetable fats. However, fat foods must not be avoided completely, but should again only be eaten occasionally.

5. Drink enough but choose your drinks reasonably
With an average of 1.6 l liquid intake in Austria is very low. Especially children and elderly people should take care of their fluid intake, but also all other persons need to drink sufficiently to maintain the body's water balance.

However, not only drinking in itself matters but also the kind of drink. Soft drinks that are rich in sugar, alcoholic beverages like beer and wine, coffee and black tea should be consumed as special enjoyments in moderation. They are not suitable to maintain the fluid balance. Instead water, diluted fruit juices or herbal teas should be preferred as there is virtually no upper limit for their intake.

Drinking reasonably means also to be careful with alcoholic drinks. Alcohol in itself is harmful, regardless of any contrary reports sometimes claimed by the media. There are almost no health indications beyond enjoyment for the consumption of alcoholic beverages, therefore they should be drunk in moderation.

6. Enjoy your foods
Daily nutrition is not a complicate affair. Observing the aforementioned guidelines takes us already a lot nearer to a healthy way of nutrition.

Delighting in one's food is as important as the sufficient supply of essential nutrients and the right amount of energy. Therefore, foods considered as unhealthy like chocolate, cream tart or fat meat are certainly allowed as is a little debauchery on festive days. The important thing is to enjoy eating these foods, and to do it only occasionally.

Accordingly, food should be chosen and handled thoughtfully, preferring high-quality products even though these might be more expensive than the mass-produced food sold in many supermarkets. Care should also be taken when preparing and serving meals.

An epicurean handling of our food helps us to eat with full awareness enjoying it with all senses and to concentrate on the taste, so that quantity is no longer the only criterion.

7. Stay active and fit with physical activity
Nutrition is not the only determinant for health. Physical activity backs a healthy way of eating and is amidst the most important factors for a better health. However, it is not sufficient to do some sports only on the weekend but to be active every day, for example by going by bicycle instead of the car, using the stairs instead of the elevator, taking a walk instead of watching TV. Physical exercise is always exhausting and overcoming idleness is not as easy. However, the great and multiple health benefits make it worthwhile, and along with practice come the pleasure in physical activity.

In general, FBDG should ensure that the nutrient needs of the population are covered and contribute in reducing the risk of chronic diseases. NBDG are a necessary starting point to develop FBDG. For the German-speaking countries (Germany, Austria, and Switzerland) NBDG exist in the form of the 'DACH Reference Values for Nutrient Intake' which are issued by the German Nutrition Society (DGE), the Austrian Nutrition Society (ÖGE), and the Swiss Society for Nutrition Research (SGE)/Swiss Nutrition Association (SVE) [4]. These standards for nutrient intake are also used to assess the diet of Austrian population groups.

Visualizations of Food-Based Dietary Guidelines in Austria

Besides the 'Healthy Eating Guidelines for Austrians', other FBDG already exist or are currently being developed in Austria. Visual aids of FBDG are essential to facilitate the promotion and dissemination of guidelines and to increase the comprehensibility to assist consumers selecting a diet that is adequate in nutrients and health-promoting. The most recognized visualizations of FBDG in Austria are the following:

Nutrition Circle and Three-Dimensional Food Pyramid
The so-called 'Nutrition Circle' ('Ernährungskreis') of the German Nutrition Society has already a long tradition in German-speaking countries. The latest revision of the Nutrition Circle happened in 2004 [5] and is based on the 'DACH Reference Values for Nutrient Intake'. The Nutrition Circle is a visual aid with seven segments representing recommended food groups which are quantitatively balanced to ensure adequate supply with all essential nutrients and to avoid overconsumption of (saturated) fat, cholesterol, salt and others.

In a next step to combine the quantitative with qualitative aspects, the Nutrition Circle has been extended to a 'Three-Dimensional Food Pyramid' [6]. Each of the four sides of the pyramid ('triangles') is allocated to particular foods: (1) foods of plant origin; (2) food of animal origin; (3) lipids and vegetable oils, and (4) beverages. In each triangle foods are categorized according their nutritional quality; foods with high nutritional quality are closer to the base whereas less desirable foods are allocated closer to the tip of the respective triangle. More details and images about these guidelines can be found in Stehle et al. [6].

The Nutrition Circle and the Three-Dimensional Food Pyramid have been adopted by the Austrian Nutrition Society and both are disseminated to health professionals and to the consumer.

Nutrition Pyramid

In 2005 the 'Fund for a Healthy Austria' (Fonds Gesundes Österreich, FGÖ) together with the Ministry of Health issued the 'Nutrition Pyramid' (Ernährungspyramide) (fig. 1). The Nutrition Pyramid is accompanied by campaigns using TV spots and advertisements in print and online media. Thus, the campaign is able to reach a broad audience. According the FGÖ, the campaign seeks to reduce risks of disease from poor nutrition and to help Austrians acquire a taste for healthier ways of eating and drinking [7].

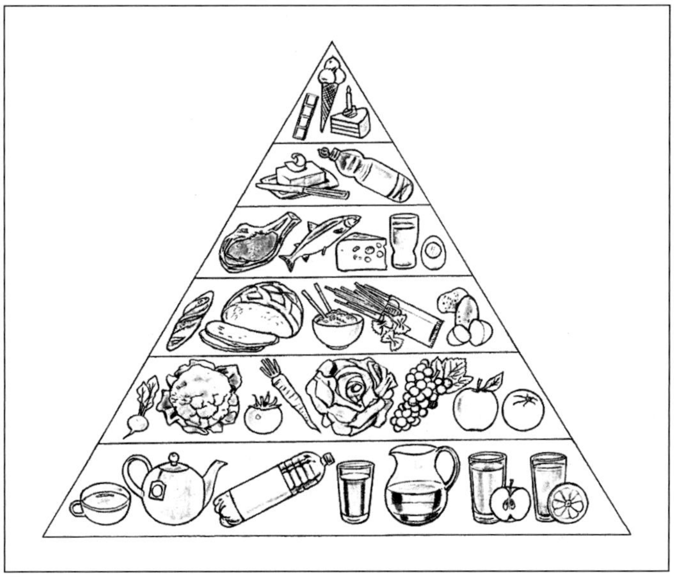

Fig. 1. Nutrition pyramid of the 'Fund for a Healthy Austria' [from 7, with permission].

The Nutrition Pyramid of the FGÖ uses the concept of the initial Food Guide Pyramid of the United States Department of Agriculture (USDA) [8, 9], which translates NBDG into a range of servings for the five major food groups (grains, fruits, vegetable, milk, and meat) to ensure an adequate supply of essential nutrients and on the other hand, restrict intake of fat, saturated fat, cholesterol, sugar, sodium, and alcohol. Unlike the USDA Food Guide Pyramid, the Nutrition Pyramid incorporates beverages at the base of the pyramid and the grain group has changed places with the fruit/vegetable group. A likely explanation for the latter is that fruits and vegetables taken together have quantitatively more weight than the grain group (as can also be seen in the Nutrition Circle). Finally, foods which should be consumed in less quantity, including lipids and sweets, have been more differentiated in the tip of the Nutrition Pyramid. The serving sizes of the different food groups follow the current recommendations of the German Nutrition Society [10], thus the Nutrition Pyramid is also based on the 'DACH Reference Values for Nutrient Intake'.

Guidelines for Cereals, Potatoes and Products Thereof: A Module System

In 2005, the Austrian Nutrition Society commenced to develop FBDG specifically for Austria taking the cultural aspects of diet in Austria into account. The guidelines follow the recommendations of the Food and Agricultural Organization (FAO)/WHO [11], the CINDI program of the WHO [12] and the EURODIET project [13].

Objectives of the guidelines

- Presentation of the scientific support for FBDG of different food groups, i.e. in a first step for cereals, potatoes and products thereof
- NBDG shall be translated into a clearly comprehensible food-based form with better practical applicability
- The DACH Reference Values for the intake of carbohydrates (at least 50% of daily energy intake) and dietary fiber (at least 30 g/day) shall be implemented by means of specific food-based recommendations for the intake of cereals, potatoes and products thereof
- Disseminators in nutritional consultation, education and information as well as food producers shall get a sound basis for argumentation. Disclosure of a balanced diet shall be facilitated by applicable and coherent recommendations
- The gap between the actual and the recommended consumption of cereals, potatoes and the products thereof shall be closed or at least reduced
- The base of a health-promoting diet shall be presented to consumers. They shall be encouraged to make a balanced food choice while still enjoying what they eat.

Aims of the guidelines

- Maintenance and improvement of health in the Austrian population
- Reduction of the incidence of nutrition-related diseases
- Applicable support in ensuring adequate nutrient supply of the Austrians
- Detailed self-explanatory instructions regarding food choice
- Encouragement to choose seasonal and regional food
- Base for further means to disseminate the Austrian nutritional guidelines.

In a first step, guidelines for cereals and potatoes ('Leitlinie für Getreide- und Kartoffelprodukte') were issued. These FBDG are quantitative guidelines for children and adults, which take respective energy and nutrient requirements into account as well as the prevention of NCDs. The corresponding quantities for the daily consumption of cereals, potatoes and products thereof are given in tables 2 and 3. The recommended amounts for adults are based on the Nutrition Circle [10], whereas for children and adolescents they are derived from the guidelines of the Research Institute for Children's Nutrition (FKE) in Dortmund, Germany ('Optimix') [14]. The amounts for each single food are given as a range according to individual energy requirements with the lower values applying to a lower energy intake and the upper values to a higher one.

The Module System: 'Take Advantage of 5 Portions'

The novelty is the visualization of these quantitative recommendations (fig. 2): a module system accompanied by the slogan *'take advantage of 5 portions' ('5 Portionen, die sich lohnen')* suggesting the consumption of five servings taken from the cereal and potato food group [15]. With this proposal the Austrian Nutrition Society offers a self-explanatory and consumer-oriented model addressed to health professionals and all stakeholders within the food chain.

One module corresponds to:

- 1–1.5 slices of bread (1 slice = approx. 50 g, so that 4 'bread blocks' supply the 4–6 slices or 200–300 g of bread recommended by the DGE)
- 1–1.5 bread rolls
- 5–6 tablespoons of muesli or cereal flakes/breakfast cereal (corresponding to about 50–60 g of rolled oats or other ready-to-eat cereals)
- 200–250 g cooked pasta (65–80 g uncooked)
- 150–180 g cooked rice or other cereal (50–60 g uncooked)
- 200–250 g potatoes (3–4 medium-sized)

Whole grain products should be preferred.

Why Five Servings?

This is based on the assumption of a daily consumption of four slices of bread (which can be partly replaced by one serving of ready-to-eat cereals) and one serving of potatoes, pasta or rice. Graphically, these servings are depicted as boxes (= modules). A text insert in each of the boxes lists alternatives corresponding to one serving. In this way the modules can be substituted for each other.

Table 2. Intake recommendations for adults for the cereal and potato group [10]

Food	Recommended intake
Bread	4–6 slices per day (200–300 g) (1 slice corresponding to approx. 50 g)
Cereal flakes, breakfast cereal	1–2 slices of bread can be substituted by cereal (breakfast) flakes. One slice is equivalent to 50–60 g of flakes
Cereal products and potatoes	One portion of potato per day (200–250 g of cooked potato or 3–4 medium-sized) or 1 serving of pasta (75–90 g uncooked, 200–250 g cooked) or 1 serving of rice or other cereal (50–60 g uncooked, 150–180 g cooked)

Table 3. Intake recommendations for children and adolescents for the cereal and potato group [14]

	Age, years						
	1	2–3	4–6	7–9	10–12	13–14, m/f	15–18, m/f
Bread/cereal (incl. flakes), g/day	80	120	170	200	250	300/250	350/280
Converted to bread slices 50 g	1.5	2.5	3.5	4	5	6/5	7/5.5
Potatoes, pasta, rice, cereals, g/day	80	100	130	150	180	250/200	280/230

Fig. 2. Pool of modules for the food group 'cereals, potatoes and products thereof' [from 15, with permission].

The following applies basically: bread and rolls should be eaten more than once a day and one serving of pasta, rice or other cooked cereal or potatoes once a day.

Booklet for Consumers

The scientific background of these guidelines as well as the module system shall be presented in an application booklet addressing the consumers after having been adapted for these target groups. Contents of this booklet will be among others:

- The importance of cereals, potatoes and products thereof for the human diet
- Explanations referring to the module system
- Information about products of the cereal and potato food group (varieties of cereal, flour and baked goods)
- Culinary hints and recipes
- Advices for the optimization of fiber and fat intake
- Information about purchase and storage
- Answers to frequently asked questions.

Moreover, the booklet for consumers shall offer practical support for health professionals in their daily counseling practice.

Conclusion and Outlook

The presented FBDG for Austria are basically consistent as they use the same concept, a nutritive approach which is based on the 'DACH Reference Values for Nutrient Intake'. Minor differences arise in the translation of this NBDG into food servings. The 'Nutrition Circle' and pyramids are the common forms used to visualize FBDG for the Austrian population. The 'Three-Dimensional Food Pyramid', which was introduced in 2005 by the German Nutrition Society and has been adopted by the Austrian Nutrition Society, combines quantitative and qualitative aspects and makes use of sophisticated dissemination tools, like a web-based virtual and interactive application. A not less promising approach has been introduced by the Austrian Nutrition Society – the 'Module System' as a visual aid to facilitate the promotion and dissemination of FBDG, probably especially to younger target groups.

However, an evaluation may be required to assess the comprehensibility and success of the presented guidelines. Altogether the introduced FBDG are meant to contribute to an improvement of nutrition of all Austrians for a better health promotion. This offers the chance to clearly reduce the incidence of many nutrition-related diseases, thus contributing to the reduction of health system costs and – probably being the most important – to an improvement of overall quality of life.

Together with the Austrian Nutrition Reports, FBDG are important measures to improve the nutritional situation in Austria. The continuation of the nutrition monitoring in Austria is in this regard the only possibility to observe changes, either positive or negative, in dietary trends and to adapt nutrition policy accordingly. Upcoming results of ongoing nutrition monitoring will evaluate the success of the current measures taken in nutrition policy.

As scientific evidence in general, food-based guidelines also underlie several developments making them not fixed for ever. Principle changes for the given guidelines, however, are not to be expected in the future. Basing on changes of the population structure it can be expected that specific guidelines for specific groups of the population have to be developed and established. This will be one of the main tasks of nutritional sciences for the coming years.

References

1 World Health Organization: Diet, Nutrition and the Prevention of Chronic Diseases. Report of a Joint WHO/FAO Expert Consultation. WHO Tech Rep Ser No 916. Geneva, WHO, 2003.

2 Elmadfa I, Freisling H, Koenig J, et al: Austrian Nutrition Report 2003, ed 1. Vienna, Institute of Nutritional Sciences, University of Vienna, 2003.

3 Elmadfa I, Burger P, Koenig J, et al: Austrian Nutrition Report 1998. Vienna, Institute of Nutritional Sciences, University of Vienna, 1998.

4 DACH – German Nutrition Society (DGE), Austrian Nutrition Society (ÖGE), Swiss Society for Nutrition Research (SGE), Swiss Nutrition Association (SVE): Reference Values for Nutrient Intake. Frankfurt/Main, Umschau Braus, 2000.

5 German Nutrition Society (DGE): Der neue DGE-Ernährungskreis. DGE Info 4/2004: 54–56.

6 Stehle P, Oberritter H, Büning-Fesel M, et al: Grafische Umsetzung von Ernährungsrichtlinien – traditionelle und neue Ansätze. Ernähr Umsch 2005;52:128–135.

7 Fund for a Healthy Austria (FGÖ): Die Ernährungspyramide baut auf. 2005. online: http://www.fgoe.org/aktivitaeten/infos-und-aufklaerung/kampagnen/archiv-kampagnen/ernaehrung-2005 [29.11.2006].

8 US Department of Agriculture/Center for Nutrition Policy and Promotion: The Food Guide Pyramid. Home Garden Bull 1992; 252

9 US Department of Health and Human Service, US Department of Agriculture: Dietary Guidelines for Americans 2005, ed 6. Washington, US Government Printing Office, 2005.

10 German Nutrition Society (DGE): DGE-Ernährungskreis – Lebensmittelmengen. DGE Info 5/2004:73.

11 FAO/WHO: Preparation and use of food-based dietary guidelines. Report of a joint FAO/WHO consultation, Nicosia, Cyprus, Nutrition Program WHO, Geneva, 1996.

12 WHO Regional Committee for Europe: CINDI dietary guide, 2000.

13 University of Crete, School of Medicine: Core report EURODIET. Nutrition & Diet for Healthy Lifestyles in Europe. Science & Policy Implications, 2000.

14 Forschungsinstitut für Kinderernährung Dortmund: optimix – Empfehlungen für die Ernährung von Kindern und Jugendlichen. Aid (ed), 2001.

15 Austrian Nutrition Society (ÖGE): Leitlinie für Getreide- und Kartoffelprodukte. Ernähr Akt 2005;2:1–16.

Annals of
Nutrition & Metabolism

Ann Nutr Metab 2007;51(suppl 2):15–20
DOI: 10.1159/000103562

Published online: July 30, 2007

Food Pyramid of the Swiss Society for Nutrition

Paul Walter Esther Infanger Pascale Mühlemann

Swiss Society for Nutrition, Bern, Switzerland

Key Words

Food-based nutrition guidelines · Swiss food pyramid · Healthy lifestyle

Abstract

The Swiss Society for Nutrition issued its Food Pyramid in 2005. It was updated according to the latest scientific evidence and is in principal agreement with food-based guidelines of other countries. It has also been officially endorsed by the Swiss government. The food pyramid stands for a balanced diet that guarantees the body a sufficient supply of energy, essential nutrients and protective substances. Food from the lower levels of the pyramid should be eaten in larger quantities, and those from the higher levels in smaller quantities. The six levels from top down represent the following food groups: sweets, salty snacks and sweetened or alcoholic drinks; oils, fats and nuts; milk, dairy products, meat, fish and eggs; whole grain products and pulses, other cereals and potatoes; fruit and vegetables; beverages. A basic principle to be communicated by the food pyramid is that there are no good or bad foods but that the relative amounts to be consumed play a key role for our health. The key message is to eat a diet that is as varied as possible and which considers foods of each pyramid level in the right amounts. The recommendations do not need to be followed every day, but should be observed on a long-term basis, e.g. a whole week. Liquid intake is an exception; liquids should be consumed on a daily basis. It is also very important to keep a healthy body weight by daily exercise as indicated by the icons on the side of the Swiss Food Pyramid.

Introduction

A special working group of the Swiss Society for Nutrition started in 2004 with the updating of the original food pyramid issued in 1998 and results from recent literature were integrated. This paper will not review the whole literature but refers to some of the more recent pertinent results. Most important were several WHO reports, especially the Global strategy on diet, physical activity and health of the 57th World Health Assembly (WHA 57.17) and its article 22.

For diet, recommendations for populations and individuals should include the following:

- Achieve energy balance and a healthy weight
- Limit energy intake from total fats and shift fat consumption away from saturated fats to unsaturated fats and towards the elimination of trans-fatty acids
- Increase consumption of fruits and vegetables, and legumes, whole grains and nuts
- Limit intake of free sugars

KARGER

Fax +41 61 306 12 34
E-Mail karger@karger.ch
www.karger.com

© 2007 S. Karger AG, Basel
0250–6807/07/0518–0015$23.50/0

Accessible online at:
www.karger.com/anm

Paul Walter
Schweizerische Gesellschaft für Ernährung (SGE)
Effingerstrasse 2, Postfach 8333
CH–3001 Bern (Switzerland)
Tel. +41 61 403 02 83, Fax +41 61 403 02 82, E-Mail paul.walter@unibas.ch

Table 1. Targets for the Swiss food pyramid

Nutrient, food components	Targets (per day)
Energy	7.4–10.6 MJ; 1,800–2,500 kcal
Carbohydrates	about 50% of total energy
Dietary fibre	about 30 g
Saccharose	moderate, about 10% of total energy
Protein	10–20% of total energy
Fat	about 30% of total energy
Long-chain saturated fatty acids	<10% of total energy
Monounsaturated fatty acids	about 10% of total energy
Polyunsaturated fatty acids	about 7% of total energy
Relation of linoleic acid to α-linolenic acid	5:1
Cholesterol	about 300 mg
Vitamins	100% DACH reference values for 19- to 65-year-old people
Minerals and trace elements	100% DACH reference values for 19- to 65-year-old people
Salt (NaCl)	about 6 g
Water from beverages and solid foods	250–270 ml/MJ or 1.0–1.1 ml/kcal (incl. water derived from oxidations)

- Limit salt (sodium) consumption from all sources and ensure that salt is iodized.

The aim of the Swiss Society for Nutrition was to present easily understandable recommendations for healthy, tasty eating and drinking for adults. Another goal was to find a consensus for a food pyramid that was accepted and endorsed by the Swiss government as well as by major Swiss organizations involved in nutritional issues including major food retailers and food industries.

For the energy as well as the macro- and micronutrient requirements, the Reference Values for Nutrition Intake DACH 2002 [1] were used. The target population for the recommendations of the Swiss Food Pyramid are healthy adults aged 19–65 with a low physical activity profile (PAL 1.4: sedentary activities, with little or no strenuous leisure activity). The average nutritive targets to be fulfilled are summarized in table 1.

Official Document of the Swiss Food Pyramid 2005 [2]

Recommendations for Healthy, Tasty Eating and Drinking for Adults

Our eating and drinking patterns are influenced by a number of different factors such as individual needs and desires, our everyday health and social environment, current food availability, advertising, etc. The following recommendations are designed for healthy adults. Other age groups and population categories (e.g. children, pregnant women, athletes, vegetarians, etc.) have specific requirements which cannot all be illustrated in the same pyramid (fig. 1).

This food pyramid shows a balanced, mixed diet that guarantees the body a sufficient supply of energy, essential nutrients and protective substances, making a major contribution to our well-being. Food from the lower levels of the pyramid should be eaten in larger quantities, and those from the higher levels in smaller quantities. All food items are allowed. The key is to eat a diet that is as varied as possible and which considers each pyramid level. This diet should be seasonal and made up of foodstuffs that have been processed and prepared with care.

A healthy diet should of course be a source of pleasure and enjoyment, which is best achieved by eating in company. The recommendations do not need to be followed every day, but should be observed on a long-term basis, e.g. a whole week. Liquid intake is an exception, as these recommendations should be followed daily.

However, our health is not merely dependent on our eating and drinking habits. To maintain a healthy body weight, it is particularly important to do at least half an hour of exercise each day (in the open air if possible). Refraining from smoking and adopting a healthy attitude towards stressful situations and/or utilising specific, conscious relaxation techniques also contribute to a healthy lifestyle.

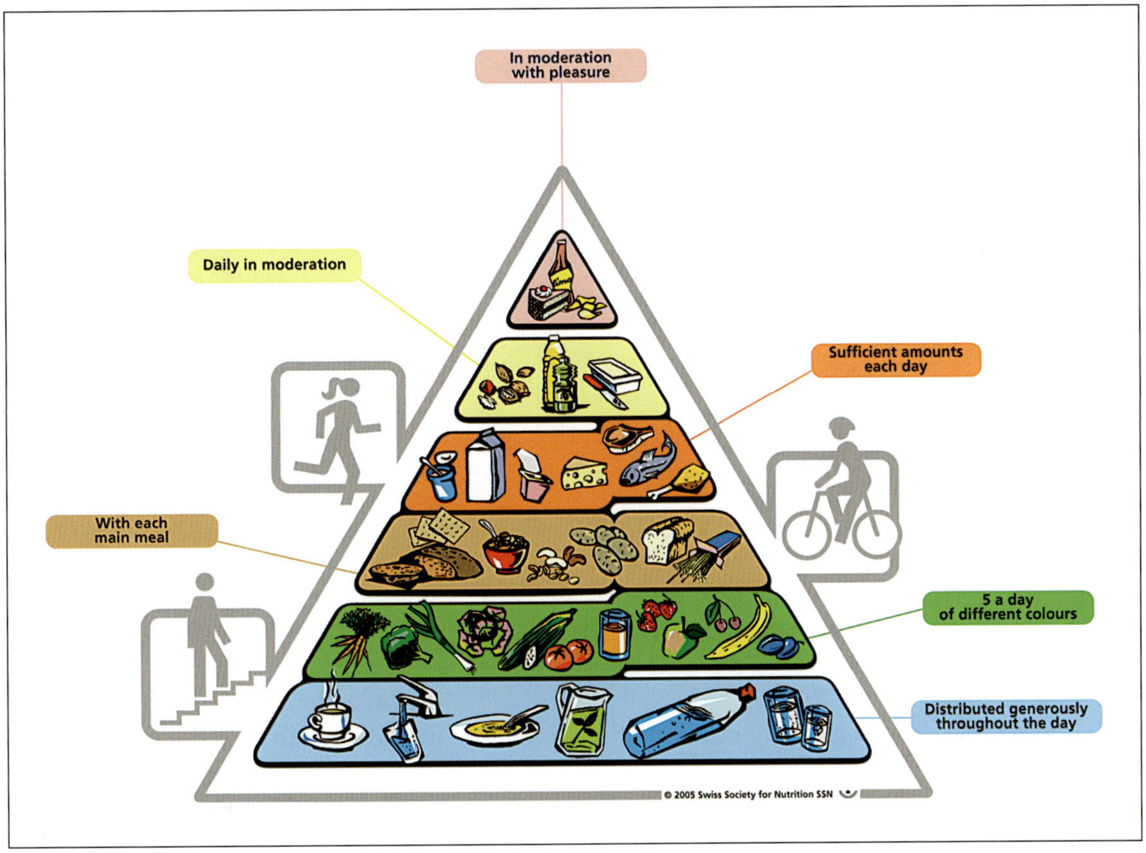

Fig. 1. Food pyramid: recommendations for healthy, tasty eating and drinking for adults.

Detailed Recommendations for the Different Levels of the Swiss Food Pyramid from Bottom to Top

Beverages – distributed generously throughout the day:
Drink 1–2 litres of liquid a day, preferably in the form of non-sugared drinks, e.g. tap/mineral water or fruit/herb teas.

Caffeine-rich beverages (coffee, black/green tea) should be drunk in moderation only.

Fruit and vegetables – 5 a day of different colours:
Eat 3 portions of vegetables a day, at least one of which should be raw (1 portion = at least 120 g of vegetables, e.g. salad, soup or as a side-dish).

Eat 2 portions of fruit a day (1 portion = at least 120 g = 1 'handful').

One daily portion of fruit or vegetables can be replaced by 200 ml of non-sugared fruit or vegetable juice.

Whole grain products and pulses, other cereals and potatoes – with each main meal:
Each main meal should be served with 1 starch-rich side-dish (i.e. 3 portions a day, 1 portion = 75–125 g of bread or 60–100 g of pulses [raw weight] for instance lentils/chick peas or 180–300 g of potatoes or 45–75 g of pasta/rice/flakes/corn/other grains [raw weight]), including at least two portions of whole grain products.

Milk, dairy products, meat, fish and eggs – sufficient amounts each day:
Each day, alternate between 1 portion of meat, fish, eggs, cheese or other sources of protein, e.g. tofu (1 portion = 100–120 g of meat/fish [fresh weight] or 2–3 eggs or 200 g of fresh/cottage cheese or 60 g of hard cheese or 100–120 g of tofu).

In addition, consume 3 portions of milk or dairy products a day, preferably reduced fat varieties (1 portion = 200 ml of milk or 150–180 g of yoghurt or 200 g of fresh/cottage cheese or 30–60 g of cheese).

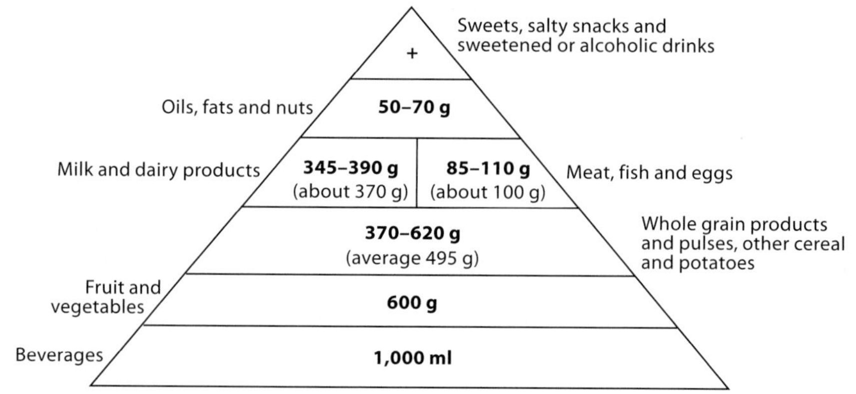

Fig. 2. Order of food groups.

Oils, fats and nuts – daily in moderation:

2–3 teaspoons (10–15 g) a day of high-quality plant-based oils such as rapeseed oil or olive oil in cold dishes (e.g. salad sauces).

2–3 teaspoons (10–15 g) a day of plant-based oils for cooking (frying, braising): olive oil is recommended for example.

Use 2 teaspoons (10 g) a day of butter or margarine made from high-quality oils to spread on bread as required.

A daily serving of 1 portion of nuts is recommended (1 portion = 20–30 g of almonds, walnuts or hazelnuts, etc.).

Sweets, salty snacks and sweetened or alcoholic drinks – in moderation with pleasure:

Consume sweets, salty snacks and sweetened drinks (e.g. soft drinks, ice tea, energy drinks) in moderation.

When consuming alcoholic beverages, do so in moderation and with a meal.

Use salt with added iodine and fluoride, but only in limited quantities.

Special Considerations

Order of Food Groups within the Pyramid

The order of the various levels of the pyramid is given by the relative amounts of the foods to be consumed. From bottom up the largest group are the beverages followed by fruit and vegetables. There is a relatively small difference of the latter group to the next one containing the main sources of carbohydrates. In special cases this may individually result in a different order of these two groups. The next level contains milk and dairy products on the one hand and other protein containing foods on the other hand (such as meat, fish, etc.). Finally there is the newly introduced level for oils, fats and nuts, and on the top of the pyramid are the 'extras' to be consumed in small amounts (sweets, salty snacks, etc.) (fig. 2).

Do These Recommendations Provide the Recommended Amounts of Nutrients?

No exact calculation can be made, since a variety of different foods can be chosen on each level of the pyramid per day (table 2). Therefore the calculations were based on a theoretical 7-day food protocol with daily changes in the choices at each level. The calculations were performed with a special software (Prodi 5 from Nutri-science). Some of the results are shown in tables 3 and 4. The macronutrient distribution, shown as percentage of total energy, as well as the distribution of the various fatty acids correspond to the recommendations (see table 1). No calculation on the ratio of ω–6 to ω–3 fatty acids could be performed. However, if the recommendation for the intake of fat and oils (especially rapeseed oil) is followed, this quotient will approach the recommended value of 5:1. Of the critical minerals measured, calcium as well as magnesium reached the recommended daily amounts, whereas zinc, iron and iodine just about reached the recommended levels. Of the water-soluble vitamins, the critical ones B_1, B_2, B_{12}, folic acid as well as vitamin C all reached the recommended levels. Somewhat surprisingly, the calculated intake of vitamin A reached only 50% and for vitamin D_3 only 20% of the recommended amounts. Due to the high availability of β-carotene in

Table 2. Calculated intake per day

3 × 120 g of vegetables and 2 × 120 g of fruit
3 × starch-rich side-dish with 30–50 g of carbohydrates each
3 × milk or dairy products with 240 mg of calcium each (all but cheese with reduced fat content)
1 × meat, fish, cheese, eggs, tofu or Quorn™ with 20 g of protein each
30–40 g oil/fat
20–30 g mixture of nuts
max. 45–60 g of sugar (= 10% of energy intake)
max. 1–2 dl of wine (approx. 10–20 g of alcohol)

Table 3. Macronutrients

Energy	9 MJ (2,150 kcal)	% of total energy
Protein	87 g	16
Carbohydrates	242 g	45
Fat	82 g	34
Dietary fibre	34 g	
Cholesterol	207 mg	
Alcohol	14.4 g	5
Beverages	1.2 l	

Table 4. Fatty acids

	Amount g	% of total fatty acids	% of total energy
Fat	82	–	34
Saturated fatty acids	26	35	12
Monounsaturated fatty acids	33	46	16
Polyunsaturated fatty acids	14	19	6
Total fatty acids	73	100	–

fruit and vegetables, enough vitamin A can be produced through cleavage of the β-carotene in the gut of the consumer. Similarly, the extra vitamin D_3 synthesis in the skin will take care of the necessary vitamin D_3 supply – which is why at least half an hour of exercise per day is recommended in the open air if possible. In conclusion, it can be said that all the critical nutrients are supplied in sufficient amounts by the recommended diet of the Swiss Food Pyramid to meet the daily requirements by DACH [1].

Special Emphasis

The main problem for full acceptance and application of the food-based guidelines by the population is the recommended high daily consumption of vegetables and fruit which are not always easily available and may result in an increase of the family budget for food. Therefore, the simpler the message the higher the chance that fruit and vegetable consumption will increase. In our experience, the food pyramid has retained its role as an excellent way to communicate the basic messages for a healthy diet in a short time.

Research of the last 10 years has fully confirmed the large beneficial effect of the 5-a-day concept of fruit and vegetable on the reduction of cardiovascular disease and stroke [3, 4] by 20 and 30%, respectively. The higher consumption of whole grain also results in a reduction of the risks for the two diseases mentioned above [5]. Even though the mechanisms for these effects are not fully elaborated, it is evident that many biologically active compounds occurring in fruit and vegetable as well as in whole grain are at least in part responsible for these effects [6]. The recommendation for a daily consumption of nuts is new; nuts have been shown to have several health effects mainly due to their healthy fat composition [7, 8].

In view of the actual obesity pandemic, the proper control of the energy balance and the correct application of the food-based guidelines by the general population can be expected to result in a strong reduction of the health costs for the treatment of the occurring diseases like diabetes and the metabolic syndrome. In Switzerland alone, these extra costs are now estimated to amount to CHF 2.5 billion per year [9]. As in many other countries it is unfortunately politically very difficult to attribute the necessary funds to implement the food-based guidelines on a large scale. The methods are available, but the finances for the implementation are at present largely insufficient.

Acknowledgements

In addition to the authors, the following individuals were members of the special working group of the Swiss Society for Nutrition: Monika Eichholzer, Priska Gnägi-Schwarz, Josef Laimbacher, Kathrin Reinli, and Paolo Colombani.

References

1 German Nutrition Society, Austrian Nutrition Society, Swiss Society for Nutrition Research, Swiss Nutrition Association: Reference Values for Nutrient Intake, ed 1. Frankfurt, Umschau Braus Publisher GmbH, 2002.

2 Swiss Food Pyramid http://www.sge-ssn.ch/science/scientific_publications/ssn_publications/food_pyramid/english/food_pyramid_65_engl.pdf

3 Dauchet L, Amouyel P, Hercberg S, Dallongeville J: Fruit and vegetable consumption and risk of coronary heart disease: a meta-analysis of cohort studies. J Nutr 2006;136:2588–2593.

4 Dauchet L, Amouyel P, Dallongeville J: Fruit and vegetable consumption and risk of stroke: a meta-analysis of cohort studies. Neurology 2005;65:1193–1197. Summary for patients in: Neurology 2005;65:E17–E18.

5 Steffen LM, Jacobs DR Jr, Stevens J, Shahar E, Carithers T, Folsom AR: Associations of whole-grain, refined-grain, and fruit and vegetable consumption with risks of all-cause mortality and incident coronary artery disease and ischemic stroke: The Atherosclerosis Risk in Communities (ARIC) Study. Am J Clin Nutr 2003;78:383–390.

6 Kris-Etherton PM, Hecker KD, Bonanome A, Coval SM, Binkoski AE, Hilpert KF, Griel AE, Etherton TD: Bioactive compounds in foods: their role in the prevention of cardiovascular disease and cancer. Am J Med 2002;113(suppl 9B):71S–88S.

7 Kris-Etherton PM, Zhao G, Binkoski AE, Coval SM, Etherton TD: The effects of nuts on coronary heart disease risk. Nutr Rev 2001;59:103–111.

8 Gebauer SK, Psota TL, Harris WS, Kris-Etherton PM: n–3 fatty acid dietary recommendations and food sources to achieve essentiality and cardiovascular benefits. Am J Clin Nutr 2006;83(6 suppl):1526S–1535S.

9 Schneider H, Schmid A: Die Kosten der Adipositas in der Schweiz. Schlussbericht für das Bundesamt für Gesundheit (BAG). Bern, BAG, 2004. www.bag.admin.ch

Annals of
Nutrition& Metabolism

Ann Nutr Metab 2007;51(suppl 2):21–25
DOI: 10.1159/000103563

Published online: July 30, 2007

Dissemination of Nutritional Knowledge in Germany – Nutrition Circle, 3D Food Pyramid and 10 Nutrition Guidelines

Peter Stehle

Department of Nutrition and Food Sciences, University Bonn and DGE, Bonn, Germany

Key Words
Wholesome diet · Nutrition circle · 3D food pyramid · German Nutrition Society · Reference values

Abstract
Background/Aims: Presentation of the general concept of the German Nutrition Society for a successful dissemination of nutritional knowledge. ***Methods:*** Evaluation of actual material used for nutrition education and teaching. ***Results:*** In Germany, different validated instruments on the basis of actual scientific evidence are available. Quantitative aspects can be easily enlightened with the help of the *Nutrition Circle*. Information concerning optimal personal behavior (lifestyle) is transported using a flyer with *10 Nutrition Guidelines*. The *3D Food Guide Pyramid* is an encouraging new instrument to establish new teaching programs with a high level of acceptance. It is planned to establish target-specific food guide pyramids, e.g. for adolescents, aged persons (considering gender), people with increased physical activity and others. Tests with target groups (health professionals, teachers, young consumers) showed a high rate of acceptance and demonstrated successful application of the German concept of dissemination of dietary guidelines. ***Conclusion:*** Using modern communication technology, it is feasible for the consumers and the health professionals to download gender- and age-specific scientific information as well as validated tools for successful nutrition education.

Introduction

Dissemination of nutritional knowledge is a valuable and widely accepted preventive measure to diminish the incidence and prevalence of chronic (non-communicable) diseases in a population. In Germany, the German Nutrition Society (Deutsche Gesellschaft für Ernährung, DGE) owns the task to establish and publish scientifically sound food-based guidelines for a balanced and wholesome diet.

In close cooperation with our sister societies in Austria and Switzerland, reference values for nutrient intake are first established [1]. These nutrient-based recommendations are then 'translated' into food-based dietary guidelines with the aim to inform the consumer how a wholesome diet should be composed as well as to teach health professionals working in the different fields of nutrition education.

The goal of the present review is to describe in detail the general procedure *from nutrients to food* and to present new educational tools for a successful dissemination of dietary guidelines.

Quantitative Aspects – the *Nutrition Circle*

According the so-called 'nutritive' approach (transfer of actual reference values for essential nutrients into food patterns) [2], the *Nutrition Circle* (fig. 1) illustrates the

KARGER
Fax +41 61 306 12 34
E-Mail karger@karger.ch
www.karger.com

0250–6807/07/0518–00021$23.50/0

Accessible online at:
www.karger.com/anm

Peter Stehle
Department of Nutrition and Food Sciences, University Bonn
Endenicher Allee 11–13, DE–53115 Bonn (Germany)
Tel. +49 228 733 680, Fax +49 228 733 217
E-Mail p.stehle@uni-bonn.de

Fig. 1. The *Nutrition Circle.*

recommended daily intake in weight percent of 7 food classes (cereals/grains, vegetables, fruits, milk products, meat/fish, oils/fats, beverages) as calculated on the basis of energy, macro-, and micronutrient needs (quantitative aspects; visualized by the respective sector area except for beverages). Considering age- and sex-specific dietary allowances (calculations made for 4 groups: men and women aged 25–51 and >65 years, respectively) and a physical activity level (PAL) of 1.4, exemplary daily plans using common foods are composed. The goal is to cover reference values for all nutrients when considering the daily intake averaged over 7 days. Food choice is strongly adjusted to the '5-a-day' principle (5 servings of fruit and vegetable per day) and the idea to use food with a high nutrient density (e.g., whole grain cereals). Food products with a known poor nutritional quality like sweets and alcoholic/energy-rich beverages are not considered.

The *Nutrition Circle* convincingly demonstrates that it is possible to provide almost all essential and important non-essential nutrients in adequate amounts by choosing high-quality food products available in supermarkets and traditional groceries. To cover iodine needs, 2 g of iodinated salt have to be consumed daily. The actual RDA for folate (folic acid) is difficult to reach even when folate-rich vegetables are consumed regularly in high amounts. In line with political measures already taken for example in USA and Canada [3], the DGE is, thus, in favor to enrich flour often used in bakeries with synthetic folic acid (150 μg/100 g) [4] to improve average intake of folate/folic acid. The low vitamin D intake from food is counteracted by its endogenous synthesis in the skin catalyzed by UVB exposition. Following the advice given in the *Nutrition Circle,* the percentage of energy from fat ranged from 28 to 31%; carbohydrates and protein provide 52–53% and 16–17%, respectively.

Indeed, the *Nutrition Circle* only considers foods with a known high physiological quality and, thus, has to be seen as a target which should be reached by the consumer.

10 guidelines of the German Nutrition Society (DGE) for a wholesome diet

German Nutrition Society
www.dge.de

A well-balanced diet keeps one healthy and promotes efficiency and well-being. The German Nutrition Society has established 10 guidelines based on most recent scientific knowledge. They will help you to stick to a healthy diet and still enjoy food.

1. Choose from among many different foods

Enjoy the great variety of food. A well-balanced diet is characterized by variety, suitable combination and adequate quantity of foods rich in nutrients and low in energy.

2. Eat plenty of cereal products and potatoes

Bread, pasta, rice, grain flakes, preferably made of whole grain, and potatoes contain nearly no fat, but plenty of vitamins, minerals, trace elements as well as dietary fibre and phytochemicals. If possible combine these products only with foods low in fat.

3. Fruit and vegetables - take "5 a day"

Eat 5 portions of fruit and vegetables a day, if possible fresh, cooked for a short time only or in form of juice - ideally along with each main meal and also in between meals: That way you'll get plenty of vitamins, minerals, dietary fibre and phytochemicals (e.g. carotenoids, flavonoids). That's the best you can do for your health.

4. Milk and dairy products daily, fish once to twice a week; meat, sausages and eggs in moderation

These foods contain essential nutrients, e.g. calcium in milk, iodine, selenium and omega-3 fatty acids in saltwater fish. Meat is favourable because of its high level of available iron and of the vitamins B1, B6 and B12. However, 300-600 g of meat and sausages per week are sufficient. You should choose low-fat products; this applies to meat and dairy products in particular.

5. Low-fat diet

Fats supply with essential fatty acids, fat soluble vitamins and energy. To much fat from food can promote overweight and probably cancer. To much saturated fatty acids promote, in the long run, cardiovascular diseases. Prefere oils and fats from plants (for example canola oil an soybean oil and margarine with this ingredients). Watch out for invisible fat in some meat products, dairy products, biscuits and sweets and also in various fast food and convenience products. Restrict your dietary fat to 70-90 g per day.

6. Sugar and salt in moderation

Food and beverages containing sugar and sugar substitutes should be consumed only occasionally. Be creative in using herbs and spices. Add little salt. At any rate use iodized salt.

7. Plenty of liquid

Water is vitally necessary. Make sure your daily liquid ration is about 1 ½ litre. Prefere water – carbonated or not carbonated - and and other beverages low in energy. Alcoholic drinks should be consumed only occasionally and if, in small amounts only.

8. Make sure your dishes are prepared gently and taste well

Cook food at low temperatures, if possible for a short time only, use little water and little fat. This will preserve the food's natural taste, protect nutrients and prevent formation of harmful compounds.

9. Take your time and enjoy eating

Eating consciously helps to eat healthy. Arrange dishes attractively. Take the time for eating. It's fun, makes you aware of the great freedom of choice and promotes the feeling of satiation.

10. Watch your weight and stay active

A balanced diet, plenty of physical activity and sports (30-60 min/d) fit together. Proper weight improves your well-being and promotes health.

Fig. 2. The *10 Nutrition Guidelines.*

Physical, Social and Technological Aspects – *10 Nutrition Guidelines*

In addition to distinct nutritive aspects, the *10 Nutrition Guidelines* (fig. 2) consider important factors of lifestyle (e.g., eating out, wellness), physical activities (recommendation to regularly perform sports), and household (e.g., food processing, storage). This one-page flyer can be easily understood and brought into practice by the consumer. In contrast to the *Nutrition Circle*, the *Nutrition Guidelines* do not only focus on the 'optimal behavior' but give more 'real' advice to follow a wholesome diet and lifestyle in daily routine.

Qualitative Aspects – the 3D Food Pyramid

Graphs are mandatory educational tools for a successful dissemination of nutritional knowledge. Pyramids, circles and other geometric forms are frequently used to help consumers in realizing nutritional guidelines [5, 6]. In recent years, the number of graphs published has increased tremendously, but unfortunately it is not always possible for the consumers to follow the rationale behind them. Presently available illustrations show obvious cons: provision of only quantitative (amount/portions of foods; see *Nutrition Circle*) or qualitative information (ranking of foods), space limitations, specific target groups (adolescents, young consumers) are not reached. Consequently, the DGE decided to establish a new graphic tool *(3D*

Fig. 3. The *3D Food Guide Pyramid,* 2006.

Food Pyramid) (fig. 3) adding important qualitative information for an adequate food choice. Qualitative ranking of food is based on the so-called 'metabolic' approach: consideration of actual data from epidemiological studies describing metabolic and clinical effects of distinct foods and/or food classes [2].

The square base of the *3D Food Pyramid* represents the *Nutrition Circle* (quantitative approach; see above) but condensed to four food classes. The four sloping sides are set to qualitatively rank products (shown as pictures/icons). On the basis of selected criteria, foods were placed on the bottom (green: preferential use), in the middle (yellow: use frequently) and on the top (red: use sparingly) of the triangle.

Foods of mainly plant origin (including cereals, bread and pastries): Major nutritive criteria for the qualitative ranking were energy density and the presence of physiologically valuable substrates like phytochemicals and dietary fiber. The glycemic index was only of minor importance. Preventive effects of distinct foods on the prevalence of chronic diseases were additionally considered. At the bottom of this side and, thus, to be consumed in larger quantities, fruits and vegetables are placed (generally low caloric, high in bioactive substrates). The second level is reserved for full bread, potatoes and low fat potato products, rice and white bread. Sweets and energy-rich snacks are at the top and should be consumed sparingly.

Foods of mainly animal origin (including meat, milk and fish products): The nutritional quality within this food class is assessed on the basis of energy density, fat content, fatty acid composition (content of essential fatty acids) and nutrient density with respect to calcium, iron, zinc, selenium, and vitamins B_1, B_2, B_6, B_{12}, and D. Similar to the evaluation of plant-derived food, preventive aspects from epidemiological studies with respect to chronic degenerative diseases were taken into account. The broad basis of the triangle is formed by fish, low-fat milk products, lean red and white meat. With decreasing physiological quality, foods like full-fat milk products/cheese, fat meat and meat products, eggs and bacon are ranked at the top.

Fats and oils: In addition to the fats delivered by foods of plant and animal origin, this food group should provide only 3% of daily total energy consumption. Decisive criteria for the qualitative assessment were the fatty acid composition (saturated, monounsaturated and polyunsaturated fatty acids; trans-fatty acids; ratio n–6:n–3) and the ratio of vitamin E to the number of double bonds present. In addition, practical aspects (use of oils/fats in cooking) as well as preventive aspects have been considered. Plant oils from canola, olives, soy, walnuts should be preferably chosen, followed by maize and sunflower oil.

Beverages: Assessment of quality was mainly based on energy density (% carbohydrates/l), content of bioactive substrates, glycemic index and caffeine content. At the bottom energy-free beverages (tap/mineral water, sugar-free fruit and herbal teas: consumption not limited) are placed. The yellow color was given to beverages with a carbohydrate concentration <70 g/l (7%; tea/coffee with added sugar, juices diluted with water, alcohol-free beer). The top (use sparingly) of the triangle is occupied by high-calorie beverages like so-called energy drinks and soft drinks. Alcoholic beverages are not considered due to the known risks for abuse.

First tests with target groups (health professionals, teachers, young consumers) showed a high rate of acceptance and demonstrated successful use of this new model in an interactive mode. In addition to computer applications and animations, two-dimensional versions ('nutrition compass') are produced as paper prints and folders.

Summary and Conclusions

In Germany, nutritional knowledge is successfully disseminated using different validated instruments which are based on actual scientific evidence. Quantitative aspects can be easily enlightened with the help of the *Nutrition Circle.* That lifestyle is an important factor to stay healthy is transported using a flyer wording *10 Nutrition Guidelines.* The *3D Food Guide Pyramid* is an encouraging new instrument to establish new teaching programs with a high level of acceptance. Within the concept it is planned to establish target-specific food guide pyramids, e.g. for adolescents, aged persons (considering gender), people with increased physical activity and others. Using modern communication technology, it will be feasible for the consumer to download an individual pyramid considering age, gender, physiological and socio-economic situation.

References

1 German Nutrition Society, Austrian Nutrition Society, Swiss Society for Nutrition Research, Swiss Nutrition Association: Reference Values for Nutrient Intake, ed 1. Frankfurt, Umschau Braus Publisher GmbH, 2002.

2 Stehle P, Oberritter H, Buening Fesel M, Heseker H: Graphics illustrating nutritional guidelines – traditional and new approaches (in German). Ernähr Umsch 2005;52:146–148.

3 Choumenkovitch SF, Selhub J, Wilson PWF: Folic acid intake from fortification in United States exceeds predictions. J Nutr 2002;132:2792–2798.

4 Krawinkel M, Broenstrup A, Bechthold A, Biesalski HK, Boeing H, Elmadfa I, Heseker H, Kroke A, Leschik-Bonnet E, Oberritter H, Stehle P: Strategies to improve folate supply in Germany – risks and benefits. Position paper of the German Nutrition Society (in German). Ernähr Umsch 2006;53:424–429.

5 Anonymous: Examples of foods in current sets of dietary guidelines and food guides; in FAO (ed): Preparation and Use of Food-Based Dietary Guidelines. Report of a Joint FAO/WHO Consultation, Nicosia, Cyprus, 1996.

6 Painter J, Rah JH, Lee YK: Comparison of international food guide pictorial representations. J Am Diet Assoc 2002;102:483–489.

Annals of
Nutrition&
Metabolism

Ann Nutr Metab 2007;51(suppl 2):26–31
DOI: 10.1159/000103564

Published online: July 30, 2007

Food-Based Dietary Guidelines in China – Practices and Problems

Keyou Ge[a] Jianbin Jia[b] Hong Liu[a]

[a]Chinese Nutrition Society and [b]Institute of Nutrition and Food Safety, Chinese Center for Disease Prevention and Control, Beijing, China

Key Words
Dietary guidelines · Food Guide Pagoda · Under- and overnutrition · Nutrition, China

Abstract
In recent decades, China has been in a rapid social and health transition. As seen in many developing countries, it is currently facing a double burden of both under- and overnutrition. The national average daily intake of energy, per reference person, was reduced from 2,328 kcal in 1992 to 2,252 kcal in 2002, but that of fat increased from 58.3 to 76.2 g. The national prevalence of growth retardation of children <5 years was 31.9% in 1992 and 14.3% in 2002. However, in these 10 years the prevalence of overweight increased by 39% and hypertension by 31% among adults. Chronic diseases induced 80.9% of the total number of deaths in 2000. Food-based dietary guidelines (FBDGs) were first developed in China in 1989, and a 'Food Guide Pagoda' was formulated in association with a thorough revision of FBDGs in 1997. They have been used as a powerful tool for mass education in dealing with under- and overnutrition. In order to inform consumers about having a rational diet, popularization of the FBDGs has been widely practiced. With the joint efforts of various sectors, about 20 million people have been involved directly. Despite the great efforts made, only a small proportion of the vast population has received the message. Governmental action is needed to spread and sustain mass education on food and nutrition in China. The Chinese FBDGs per se need to be revised and specific guidelines for different populations such as infants and elderly people should be developed.

Introduction

Food-based dietary guidelines (FBDGs) are widely used for putting into practice the nutrition education goals in both developing and developed countries. Departing from the dietary guidelines commonly applied in the past which were more based on nutrient requirements and food groups, FBDGs promote mainly locally available foods and traditional cuisines, emphasize local adaptation and application. FBDGs should be culturally sensitive and related directly to the communities in question. Not only should current nutritional problems be taken into account, but also those predicted for the future.

While undernutrition such as protein-energy malnutrition, deficiencies in vitamin A, iodine and iron is still the most important nutritional problem in some parts of the world, there has been a dramatic increase in the prevalence of chronic non-communicable diseases (CNCDs), such as overweight/obesity, diabetes, hypertension, etc. Many developing countries including China are today suffering from a double burden of both under- and overnutrition. Following the International Conference of Nu-

KARGER
Fax +41 61 306 12 34
E-Mail karger@karger.ch
www.karger.com

0250–6807/07/0518–0026$23.50/0

Accessible online at:
www.karger.com/anm

Keyou Ge
President, Chinese Nutrition Society
6# Guang-An-Men-Nei St. Fenghua Square, A-5-1601
Beijing 100053 (China)
Tel. +86 10 8355 4787, Fax +86 10 8355 4780, E-Mail gky@cnsoc.org

Table 1. Trends of food consumption in 20 years in China (g/day/reference person)

	National			Urban			Rural		
	1982	1992	2002	1982	1992	2002	1982	1992	2002
Cereals	509.7	439.9	401.7	459.0	405.4	366.1	531.0	485.8	415.3
Beans and nuts	17.4	14.3	19.9	17.8	16.7	20.9	14.7	13.2	19.5
Vegetables and fruits	367.5	369.2	331.2	382.4	407.4	329.6	361.2	349.5	331.7
Animal sources	60.7	117.3	159.5	109.0	210.2	248.3	40.2	69.4	125.4
Visible oil and fat	18.2	29.5	41.4	25.8	36.9	44.0	14.9	25.6	40.4
Salt	12.7	13.9	12.0	11.4	13.3	10.9	13.2	13.9	12.4

trition in 1992, that was jointly organized by the WHO and FAO, a number of governments formulated the country's National Plan of Action for Nutrition. In many Asian countries the dual problems of under- and overnutrition are addressed, recognizing the long-existing problems of stunting and underweight in children, and also the trend of increasing prevalence of CNCDs. The FBDGs are commonly recommended as a useful tool in dealing with these conditions. In addition to diet, other risk factors, such as inactivity, stress, tobacco and alcohol use, are also contributing to the changing health status. Therefore, physical activity and a healthy lifestyle have commonly been advocated in recent decades in association with the promotion of a balanced diet.

The Chinese Nutrition Society formulated the first FBDGs in 1987; a thorough revision was made in 1997 and meanwhile the Society has formulated a 'Food Guide Pagoda' to illustrate the food groups and the amount of each group to be consumed every day. After 10 years of practice, we have accumulated some experience in using the FBDGs as a tool for mass education and learned some lessons at the same time. A new nationwide nutrition survey which was completed in 2002 provided new dietary and health information for the Chinese, and suggested a further revision of the Chinese Dietary Guidelines.

Rapid Social and Nutritional Transition in China

China's gross domestic product (GDP) has grown rapidly in recent decades. The GDP has risen from RMB 454.56 billion in 1980 to RMB 1,866.78 billion in 1990, RMB 9,921.46 billion in 2000 and RMB 18,308.48 billion in 2005. The average per capita GDP was RMB 463, 1,644, 7,858 and 14,040 respectively in these years. The indices of GDP, calculated at constant prices, were 116.0 in 1980, 282.5 in 1990, 705.6 in 2000 and 1204.4 in 2005 [1].

In association with the economic growth, notable dietary changes have also occurred in this period. A dietary survey which covered 68,962 subjects of 23,470 households (7,687 urban, 15,783 rural) was conducted in 2002 [2]. According to the News Office of the State Council in 2004 [3], the national average consumption of cereals was 408 g/day per reference person, about 100 g less than that observed in 1982; however, the consumption of foods of animal sources were about 100 g more and visible oil more than doubled in these 20 years. The consumption of vegetable and fruit reduced slightly in both urban and rural areas (table 1).

Nutrients intake has changed accordingly during this period of time. Average daily energy intake reduced from 10,424 kJ (2,491 kcal) per reference person in 1982 to 9,416 kJ (2,251 kcal) in 2002. Protein intake remains mainly at the same level, and dietary fat increased from 68.3 to 85.5 g in the urban population and from 39.6 to 72.2 g in the rural population. Dietary fat provided 35 and 28% of the energy for the urban and rural population, respectively [4]. The national average of energy from fat was 30% in 2002, reaching the upper limit of a rational diet for sedentary individuals as recommended by a Joint Expert Consultation organized by the WHO and FAO of the United Nations in 1993 [5].

Anthropometric measurements of 7,026 urban and 8,825 rural preschool children indicated a reduction of undernutrition in the last 10 years. The prevalence rate of stunting among urban children was 19.1% in 1992 and 4.9% in 2002; that among rural children was 35.0 and 17.3%, respectively (by WHO/NCHS reference). The prevalence of underweight was reduced from 10.1 to 3.1% in urban children and from 20.0 to 9.3% in rural children in these 10 years. Undernutrition in adults (BMI <18.5 kg/m^2) reduced slightly from 8.3 to 7.6% in urban areas and from 10.2 to 8.9% in rural areas in the last 10 years [6].

Table 2. Average height, weight and BMI of Chinese adults measured in 1992 and 2002

		Sample		Height, cm		Weight, kg		BMI, kg/m^2	
		1992	2002	1992	2002	1992	2002	1992	2002
Urban									
Males	18–59 years	6,918	15,483	167.6	168.1	63.29	67.07	22.52	23.73
	60+ years	1,466	5,923	164.4	164.8	62.81	65.4	23.24	24.08
Females	18–59 years	8,228	22,397	156.3	156.3	55.9	57.85	22.87	23.67
	60+ years	1,472	6,297	152.1	152.6	55.05	57.5	23.80	24.69
Rural									
Males	18–59 years	14,595	35,854	164.9	165.9	58.45	61.92	21.50	22.50
	60+ years	2,168	7,549	161.7	162.1	54.71	57.3	20.93	21.80
Females	18–59 years	16,710	46,686	154.3	155.0	52.06	54.76	21.86	22.79
	60+ years	2,043	7,261	149.5	150.1	47.00	50.4	21.02	22.37

Average adult height was found to be 0.6–1.0 cm higher for rural groups in the 2002 survey than those in the 1992 survey, but remained almost unchanged for the urban groups, i.e. it increased no more than 0.5 cm. On the other hand, average body weight was 2.6–3.5 kg more in rural groups and 1.9–3.8 kg more in urban groups in 2002 than in 1992. As a result, the average BMI of Chinese adults has risen by 0.80–1.35 kg/m^2 (table 2).

A group of scholars undertook an extensive review of Chinese research studies and recommended a set of cutoffs for categorization of overweight and obesity for adults in 2004 [7]. It proposed BMI ≤24 and BMI ≥28 for overweight and obesity respectively. Among a total of 140,022 subjects (≥18 years) measured in the 2002 survey, the prevalence of overweight was 31.1% in males and 22.7% in females in the urban group, and 19.6% in males and 21.4% in females in the rural group. 10.3% of the males and 9.5% of the females in the urban group and 4.9% of the males and 6.8% of the females in the rural group were suffering from obesity in 2002; the national average prevalence of overweight was 22.8% and obesity was 7.1% based on these cutoffs. However, categorized with the WHO reference (BMI ≥25 and BMI ≥30 as cutoffs), the prevalence of overweight and obesity was averaged 18.9 and 2.9% among Chinese adults [8].

The morbidity of CNCDs has increased rapidly in line with the economic growth. Based on the 2002 China Nutrition and Health Survey, the prevalence of hypertension in people >18 years of age was 18.8%, and it is estimated that >160 million people are suffering from this illness. Compared with that found in the 1991 survey, the prevalence of hypertension increased by 31%, which means there are >70 million new patients. The prevalence of hypertension in rural areas has also increased rapidly and there is no significant difference between urban and rural populations. The prevalence of hypertension in big cities and medium to small cities was 20.4 and 18.8%, respectively, and that in four different classes of rural areas was from 12.6 to 21.0%.

The prevalence of type 2 diabetes in people >18 years of age was 2.6%, and that of impaired fasting blood glucose was 1.9%. It is estimated that there are >20 million diabetic patients plus nearly 20 million people with an impaired fasting blood glucose level in China. Compared with the result of the 1996 sample survey, the prevalence has increased from 4.6 to 6.4% in big cities and from 3.4 to 3.9% in medium and small cities. The prevalence of diabetes, different from that of hypertension, is significantly higher in urban areas than in rural areas, and is significantly higher in relatively affluent rural areas than in poor rural areas. The prevalence of abnormal blood lipid levels was 18.6% among the adults, affecting about 160 million people [3]. Chronic diseases have mostly contributed to deaths in China in recent years. Based on the statistics of the Ministry of Health, chronic diseases induced 80.9% of the total number of deaths in 2000 [unpubl. data].

Formulation of the Chinese Dietary Guidelines and the Food Guide Pagoda

The first Chinese dietary guidelines, formulated in 1989 by a group of senior nutritionists, were recognized by the Standing Committee of the Chinese Nutrition Society and published in the same year. It was named 'China's Dietary Guidelines' and consisted of the following eight items: (1) Eat a variety of foods; (2) Do not eat too much; (3) Consume the right amount of fat and oil; (4) Coarse foods work appropriately in fine foods; (5) Limit the amount of salt used; (6) Eat less sweetmeats; (7) Drink advisably, if you drink alcohol, and (8) Rationally arrange the daily three meals.

Each item of these guidelines consists of the same number of Chinese characters and can be read in sounds of some cadence. Therefore, they were transmitted widely in China in the recent decades [9].

During the 1990s, the accumulation of scientific evidence emphasized strongly the importance of a combination of a rational diet with appropriate physical activity and healthy lifestyle in health promotion. The advance of FBDGs in both developed and developing countries urged a revision of 'China's Dietary Guidelines' [10]. On the other hand, the diet of Chinese people has changed with remarkable defects during this period of time, e.g. too little amounts of milk and beans have been consumed. Aimed at tackling the practical problems in Chinese diet, new FBDGs are to be formulated.

The 'Dietary Guidelines for Chinese Residents' was prepared by a Commission composed of experts from The Chinese Nutrition Society and The National Institute of Nutrition and Food Hygiene, affiliated with the National Center for Diseases Prevention and Control, with the commission of the Ministry of Health, China. The Guidelines are a thorough revision of the China Dietary Guidelines released in 1989. The new version of Guidelines was recognized and officially released in association with the Food Guide Pagoda by the Standing Board of the Chinese Nutrition Society in 1997 [11].

The Guidelines are founded on principles of nutritional science and the present health situation of the Chinese, taking into account the traditional dietary practice of the population and the main problems currently existing in Chinese diet. The new guidelines also consist of 8 items and are aimed at providing guidance to the Chinese people for selecting a balanced diet so as to enjoy a rational nutrition and good health. They are: (1) Eat a variety of foods, with cereals as the staple; (2) Consume plenty of vegetables, fruits and tubers; (3) Consume milk, beans or dairy products, bean products every day; (4) Consume appropriate amounts of fish, poultry, eggs and lean meat; reduce fatty meat and animal fat in the diet; (5) Balance food intake with physical activity to maintain a healthy body weight; (6) Choose a light diet that is also low in salt; (7) If you drink alcoholic beverages, do so in limited amounts, and (8) Avoid unsanitary and spoiled foods.

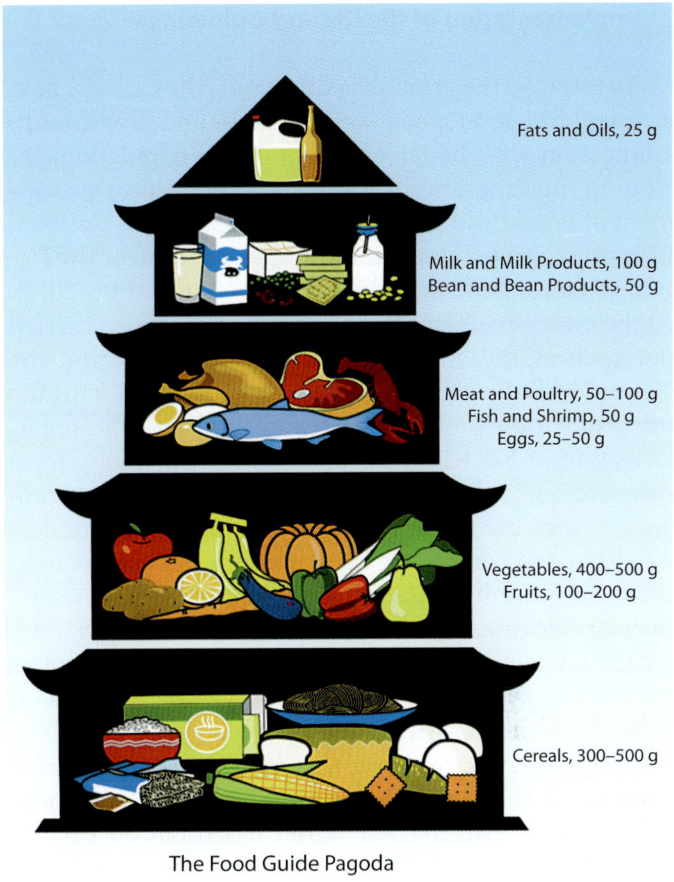

Fig. 1. Food Guide Pagoda, China.

To help consumers put the Guidelines into practice, the Commission proposed a daily food recommendation and presented it in the form of a Food Guide Pagoda (fig. 1). The pagoda visually shows the concept of food classification and the recommended daily intake of each of the five food groups. It visually shows consumers what foods and how much they should eat every day.

To take care of the specific nutritional requirements and physical and behavioral characteristics of the specific groups of people, the Expert Committee also developed the 'Recommendations for Particular Groups of People' as a supplement to the Guidelines for the general population. It covers seven population groups, namely infants, toddlers and preschool children, school-age children, adolescents (teenagers), pregnant women, lactating mothers, and elderly people. A few concise recommendations were proposed for each group.

Implementation of the Dietary Guidelines

To improve the general health status, the Chinese government has developed a series of plans and programs in connection with better nutrition of the population, including the Program for National Reform and Development of Food Production during the 1990s; the National Plan of Action for Nutrition, and the Program for the Development of Food and Nutrition in China (2001–2010). At the same time, a number of activities have been carried out such as the 'Soybean Campaign' in 1996 and the 'School Milk Program' in 1999, to mainly deal with the undernutrition problems of schoolchildren. The popularization of FBDGs is considered as being a long-term measure for achieving the national overall nutrition goals, which are not only aimed at reducing both undernutrition and controlling overnutrition, but also focused more on preventing CNCDs, e.g. obesity, diabetes, cardiovascular disease, osteoporosis and some cancers. After the guidelines were published, a number of activities for their implementation took place in various ways on different scales in different regions.

(1) *Trial of effectiveness of the guidelines as a mass education tool:* Soon after the publication of the guidelines, trials of their effectiveness were carried out on various scales. Zhao et al. [12] studied 5,145 subjects selected from five cities with different geographical and economical conditions. After repetitious promotions illustrating the guidelines and pagoda while people were gathered together, the proportion of people who understood the meaning of the guidelines grew on average from 12 to 93% within 1 year. The effect was more notable among schoolchildren and elderly people. Nutrition knowledge examinations showed that 48–59% of the people answered the questions correctly before the intervention and 68–91% after the intervention. Meanwhile, dietary practice had changed and become healthier. 52.5% of the schoolchildren had a relatively balanced breakfast after the intervention compared with 26% before; the proportion of students drinking soft drinks declined from 22.5 to 10.8%, indicating that more than half cola drinkers had switched to water in school. Among the elderly people, smokers had reduced by 16.1% points. Similar results were also reported by a number of investigators of some provincial institutions, such as Zhang et al. from the Provincial Nutrition Society of Shandong Province, and Liu et al. from the Institute of Sanitation and Disease Prevention of Sichuan Province, etc. [unpub. data].

(2) *Transmitting knowledge on food and nutrition to the public:* As a national scientific social group, the Chinese Nutrition Society has made active efforts in promoting mass education. Multi-measures have been used where appropriate to transmit balanced knowledge of food and nutrition to the public, such as lectures, training, television programs, newspapers, magazines, posters, throwaways and CDs, etc. The council members of the Chinese Nutrition Society, most of whom are leaders of local nutrition societies at provincial or municipal levels, plied a prolonged very active role in disseminating knowledge to the public. Taking the Shanghai Nutrition Society as an example, they organized 312 lecture courses explaining the FBDGs to the public in 18 months, and about 36,600 audiences joined the lectures. The lectures were always given by senior persons and recognized highly by the participants. At the same time, they published 153 tractates in newspapers, 61 television shows, 12 street poster exhibitions and mass consultations. A total of about half a million pamphlets were distributed. Similar activities were also carried out in Guangzhou, Tianjin and other cities (work reports of local nutrition societies [unpubl. data]).

The Gaps and Future Tasks

(1) Despite the great effort that has been put into spreading balanced knowledge about food and nutrition, the voice is still too weak to reach more consumers. Television advertisements on food and functional food are overwhelming, much stronger than the scientific voice, since there is no governmental budget for mass education on food and nutrition in China yet. Therefore, only a small proportion of the vast Chinese population knows about FBDGs. To overcome this difficulty, governmental actions are urgently required.

(2) In addition to nutrition-related chronic diseases, behavioral and psychological disorders, cognitive impairment, environmental diseases and risks of new foods, etc., have recently been strongly attracting the attention of the government, the professional community and the general public. To minimize these new health problems, FBDGs have become more challenging in the field of nutrition. We believe that the current 'Dietary Guidelines for Chinese Population' seem to be a little out of date and need a timely revision. The Chinese Nutrition Society is organizing a task force to investigate the situation this year and to make a revision of the Chinese dietary guidelines in 2007.

The new version of the Chinese FBDGs will take into account the recent advances in food and nutrition research works, and also consider the current situation of nutrition and health status of the Chinese population such as:

(1) The status of nutrition and health among the Chinese people has been significantly improved. Energy and protein intakes are basically satisfactory and the food diversity and dietary patterns have been improved among both urban and rural populations. The average weight of newborns has reached 3,309 g, a low birth weight accounts for 3.6% of the total number live births; the prevalence of stunting and underweight among preschool children has been decreased by more than half in 10 years; the prevalence of anemia among Chinese people has also decreased slightly.

(2) Nutrition and health problems in the population should not be ignored. The dietary pattern among the urban population needed an intervention since the consumption of animal products was too high, and the proportion of energy from cereals was too low; malnutrition among children is still quite serious in some rural areas, particularly in Western China; micronutrients deficiency such as iron and vitamin A deficiencies are rather common among urban and rural populations; the rapid increase in morbidity of hypertension, diabetes, obesity and abnormal blood lipid levels all require closer attention.

The new version of the Chinese Dietary Guidelines will possibly be issued in various types of layout to meet the needs of different population groups, possibly also as a complete version and a simplified booklet. It may emphasize more on physical activities and on a healthy lifestyle in addition to a balanced diet. It may include more detailed guidance for specific groups such as pregnant women, children and elderly people, etc. In general, the people's dietary practice should maintain the traditional food pattern whilst reducing pork consumption, reducing the amount of cooking oil used and the total fat intakes. Nutrition policies should be coordinated with other policies, such as population control and agricultural production policies.

In association with the revision of the new guidelines, we are planning to strengthen the training of nutrition workers in China for the implementation of the FBDGs. This will include the development of training materials and conducting a short-term training course. The training material will emphasize the application of the FBDGs based on local conditions, and also emphasize the promotion of related training programs in cooperation with appropriate local organizations, particularly the relevant government institutions. The training course will conducted in 28 training bases established in cooperation with the local nutrition societies located in the provinces and municipalities. The aim of all this work is to promote, as far as possible, the popularization of FBDGs in China as effectively as possible in the coming years.

References

1 National Bureau of Statistics of China: China Statistical Yearbook. Beijing, China Statistics Press, 2006, pp 56–60.
2 Zhai FY, He YN, Wang ZH, Yu WT, Yang XG: The status and trends of dietary nutrients intake of Chinese population. Acta Nutr Sin 2005;27:181–184.
3 Ministry of Health, Ministry of Science and Technology, National Statistical Bureau, China: The Status of Nutrition and Health of Chinese inhabitants, Material for Press Congress of the News Office, State Council, Oct 12, 2004.
4 Jai FY, Yang XG: Dietary and Nutrient Intakes – 2002 Nutrition and Health Survey of Chinese Residents. Report II. Beijing, People's Medical Publishing House, 2006, pp 58–67.
5 FAO Food and Nutrition Paper 57: Fats and Oils in Human Nutrition – Report of a Joint Expert Consultation. Rome, FAO, 1994, pp 1–4.
6 Yang XG, Zhai FY: Physical and Nutritional Status of the Population – 2002 Nutrition and Health Survey of Chinese residents. Report III. Beijing, People's Medical Publishing House, 2006, pp 1–157.
7 Cooperation of Overall Data Analysis, the Working Group on Obesity Problems in China: Predictive values of body mass index and waist circumference to risk factors of related diseases in the Chinese adult population. Chin J Epidemiol 2002;23:5–10.
8 Wang LD: Overview report – 2002 Nutrition and Health Survey of Chinese residents. Report I. Beijing, People's Medical Publishing House, 2005, pp 50–52.
9 Li JS: The introduction; in Expert Committee of the Dietary Guidelines for Chinese Residents (ed): A Corpus of Dietary Guidelines for Chinese Residents. Beijing, Jiancha Publishing Co, 1999, pp 1–7.
10 Regional Office for the Western-Pacific, WHO: Development of Food-Based Dietary Guidelines for the Western Pacific Region. Manila, WHO, 1999. ISBN 92-0961-129-4.
11 Ge K, McNutt K: Publication of the Chinese dietary guidelines and pagoda: How it happened here. Nutr Today 1999;34:104–115.
12 Zhao LY, Zhai FY, Li D, Li Y, Chang SY, Hao HF, Jia FM: A survey on the effects of 'Dietary Guideline' nutrition education project. J Hyg Res 2001.30.176–179.

Annals of Nutrition & Metabolism

Ann Nutr Metab 2007;51(suppl 2):32–35
DOI: 10.1159/000103565

Published online: July 30, 2007

Development of Food-Based Dietary Guidelines for Iran: A Preliminary Report

S.M. Safavi[a] N. Omidvar[b] A. Djazayery[a] M. Minaie[a] A. Hooshiarrad[c] R. Sheikoleslam[a]

[a] The Community Nutrition Department of the Ministry of Health and Medical Education, Tehran, Iran; [b] Department of Community Nutrition, School of Nutrition Sciences and Food Technology, Tehran, Iran, and Iranian Nutrition Society (ATA); [c] Department of Nutrition Research, National Nutrition and Food Technology Research Institute, Tehran, Iran

Key Words

FBDGs · Iran, development of FBDGs · Nutrition education

Abstract

Iran is undergoing an epidemiological and nutritional transition. Over the last 2–3 decades, major changes in lifestyle and food consumption patterns have occurred, as well as in the socio-economic situation. Disease patterns have changed and some micronutrient deficiencies and chronic nutritional diseases, including obesity, are on the increase. Considering these, food habits, and the national food system, the Nutrition Department, Ministry of Health and Medical Education and the Iranian Nutrition Society jointly undertook to prepare food-based dietary guidelines (FBDGs). The process consisted of three phases: agreeing on food groups, designing pictorial FBDGs, and field testing them. Three working teams, supported by a large group of advisors consisting of nutritionists, epidemiologists, food/nutrition planners, and other relevant experts were formed. The first draft of the food groups consisted of 7 groups: Bread and Cereals; Meats and Eggs; Pulses; Fruits; Vegetables; Milk and Dairy Products; Miscellaneous (nuts, oils and fats, sugar and sweets, soft drinks). It was sent to 20 nutrition and public health experts for feedback. A suggestion was made to divide the Miscellaneous group into two – oils and fats, and Miscellaneous (sugars and sweets, pickled and salted food items). Three pictorial FBDGs were then designed: a plate, a pyramid, and a rectangle. Thus, three posters were prepared by the working teams and revised by a group of experts, followed by a preliminary evaluation. The final step will be field testing of the three revised, modified designs by the public and experts and their final evaluation.

Introduction

The history of developing food-based dietary guidelines (FBDGs) in Iran dates back to the early 1960s, when the National Food and Nutrition Institute presented a poster showing the Four Basic Food Groups. In 1990, dietary guidelines, basically centered on food diversity, better methods of home food preparation, low salt and sugar intakes, breastfeeding, weight watching, and prevention of obesity were proposed and announced by the Iranian Nutrition Society [1]. The Society further revised the national food groups and dietary guidelines and included messages about physical activity. Recommendations were also made to adopt appropriate strategies to promote the nutritional awareness of the public.

Basically, food groups and dietary guidelines for any country – or even a region in a country – should be designed with due consideration of the prevailing health

KARGER

Fax +41 61 306 12 34
E-Mail karger@karger.ch
www.karger.com

0250–6807/07/0518–0032$23.50/0

Accessible online at:
www.karger.com/anm

N. Omidvar, Department of Community Nutrition
School of Nutrition Sciences and Food Technology
Shahid-Beheshti University of Medical Sciences
PO Box 19395/4741, Tehran (Iran)
Tel. +98 21 2236 0661, Fax +98 21 2236 0660, E-Mail nomidvar@yahoo.com

and nutrition situation, as well as of the socio-economic profile. Thus, before describing the process of developing the food groups and dietary guidelines, it will be appropriate to discuss briefly the health and nutrition situation in Iran.

As a result of lifestyle, food patterns, and socio-economic changes over the recent decades, Iran is undergoing nutrition transition [2, 3]. Furthermore, successful health policies and programs have helped significantly in an increase in the life expectancy and promotion of public health. Disease patterns have changed. Studies show that diseases, such as diabetes mellitus, cancers, cardiovascular diseases, and obesity, have an increasing trend, while infectious diseases, although under control, are still problems. Another nutritional problem of great concern is micronutrient deficiencies. Sections of the population suffer from a deficiency, or at least a low intake of iron, vitamins A and D, riboflavin, and calcium. Wasting and stunting are seen in some parts of the country [4]. On the other hand, all age/sex groups are at the risk of overweight and obesity due to a low level of regular physical activity and unsound food habits and nutritional behavior [5]. National food consumption surveys reveal that the staple food is cereals – mainly bread and rice. On the basis of the Food and Agriculture Organization and World Health Organization (FAO/WHO) [6] and the United States Department of Agriculture (USDA) [7] recommendations, the average daily intakes of animal products and pulses are low [8], oils and fats and simple sugar consumptions are on the increase, and the trend of animal products and pulses consumption is a downward one.

On the average, more than 50% of the daily energy and protein intakes are supplied by cereals [8]. The shares of animal products – which provide 35% of the daily intake of calcium and 55% of that of vitamin A – in supplying energy and protein are only 12 and 30%, respectively. The mean per capita daily energy intake in both urban and rural areas exceeds 100%.

The Process

The Nutrition Department of the Ministry of Health and Medical Education and the Iranian Nutrition Society (ATA) decided to prepare a food group plan and FBDGs for the Iranian population. In order to proceed systematically, groups of individuals with a wide range of expertise were formed, each charged with a specific task, with technical advice and support from nutritionists, food/agriculture scientists, epidemiologists, planners, and other related experts.

Firstly, information was collected on food consumption patterns, common nutrition-related diseases, cultural characteristics, and ease of access to foods in the community. Then a draft food group plan was designed and proposed with due consideration of the above factors and variables. The draft consisting of 7 food groups (Bread and Cereals, Meats and Eggs, Pulses, Vegetables, Fruits, Milk and Milk Products, and Miscellaneous) was commented by a group of food, nutrition and other experts. The comments and views were discussed in a meeting and changes were made. The revised draft was sent to another selected group of experts, followed by individual structured interviews with them. The final food group plan proposed by the team was as follows:

Group 1 – Bread and Cereals: all types of bread, rice, macaroni, wheat, barley

Group 2 – Meats and Eggs: all types of red meat, all types of white meat, eggs, shrimps

Group 3 – Pulses: all types of beans and peas, nuts

Group 4 – Vegetables: all types of leafy vegetables, all types of non-leafy vegetables

Group 5 – Fruits: apples, citrus fruit, grapes, other fruits

Group 6 – Milk and Milk Products: milk, yoghurt, cheese (other local dairy products)

Group 7 – Fats and Oils: all types of fats and oils, with emphasis on non-hydrogenated oils

Group 8 – Miscellaneous: sugars and sweets, salted and pickled products.

The next activity was development of FBDGs in the form of statements/sentences after thorough discussions and exchange of ideas. The FBDGs included statements/sentences about the following topics: Selecting a balanced diet; maintenance of ideal body weight; regular exercise; increasing fiber intake through eating high-fiber food items; increasing intake of low-fat milk and milk products; decreasing solid fat intake by eating less fried foods; eating white meats, in preference to red meat, as far as possible; reducing salt and sugar intakes.

In the next step, it was decided to develop pictorial food groups and FBDGs. Literature search showed several pictorial representations of food groups and FBDGs – different with regard to color and design – currently in use in different countries. An important factor in deciding what design and color to use is evidently the cultural characteristics of a population/community [9]. The most commonly used designs were found to be the food pyramid [10], the pagoda (traditional multistorey

Table 1. Food groups for the Iranian population, 2002

Group 1	Bread and Cereals: all types of bread, rice, pasta, wheat, barley
Group 2	Vegetables: all types of leafy vegetables, and non-leafy vegetables
Group 3	Fruits: apples, citrus fruit, grapes, other fruits
Group 4	Milk and Milk Products: milk, yoghurt, cheese, other local dairy products
Group 5	Meats, Eggs, and Pulses (legumes): all types of red meat, all types of white meat, eggs, shrimps, all types of pulses, nuts
Group 6	Miscellaneous: fats and oils, sugars and sweets, salted and pickled foods

Table 2. FBDGs for the Iranian population, 2002

The ABC of Healthy Nutrition:
▷ Sound selection of foods
▷ Sound diet planning
▷ Physical activity

1 Select a balanced, varied diet
2 Eat in amounts so that your ideal body weight will be maintained
3 Try to exercise regularly, at least 3 days per week, 30–40 min each time
4 Through eating more whole wheat bread, fruits, vegetables, and legumes, increase your fiber intake
5 Try to increase your intake of low-fat milk and milk products
6 Reduce, as much possible, your solid fat and fried food intake
7 Eat white meats as far as possible
8 Reduce your intake of sugar
9 Reduce your overall intake

house in the Far East) [11, 12], the food circle [9, 13, 14], the plate [15], and the rainbow [16]. After thorough discussions, exchange of ideas, and technical consultations, and bearing in mind the most common cultural characteristics, pictorial food groups and FBDGs were designed to be field tested subsequently. The designers were professional ones briefed by the consultants. Agreement was reached on three designs – pyramid, plate, and rectangle – on the basis of which three posters were prepared.

This was followed by a preliminary evaluation and field testing of the food group plans and FBDGs aiming at determining their acceptability and understandability by representative samples of three groups of health and nutrition experts, health workers, and the public (mothers and local female health volunteers). Focus-group discussions and seminars were held for this purpose in several locations. Three posters were prepared based on the three designs agreed to – the pyramid, the plate, and the rectangle – and presented in actual sizes, and discussions were held. In addition, slide shows were given and the views and comments of the participants recorded. The final contents of the food group plan and the pictorial designs were prepared after analysis of the comments and views of the three groups. The pyramid was the design selected by the majority of the participants, followed by the plate and the rectangle.

The Final Results

The final food groups and FBDGs for the Iranian population are presented in tables 1 and 2, respectively. A final evaluation of the pictorial designs will be made subsequently before official declaration. It was agreed unanimously that the process of preparing the FBDGs is a dynamic process and that, therefore, continuous evaluation and revision is essential. It is recommended that these guidelines be revised every 5 years.

Acknowledgements

This project was funded by the World Health Organization Representative's Office in Tehran, Iran. Valuable contribution of the members of the Iranian Nutrition Society (ATA) and the Community Nutrition Department of the Ministry of Health and Medical Education, Tehran, in the conduct of the project is greatly acknowledged. Thanks are also extended to all the experts who participated in the focus-group discussions and meetings. We would also like to express our deep appreciation to all the health workers and mothers in the Health Centers who cooperated closely in this project.

References

1 Resolution of the First Iranian Nutrition Congress on The National Nutritional Problems, December 26–30, 1990, Tehran. Tehran, National Nutrition and Food Technology Research Institute, 1991.
2 Ghassemi H, Harrison G, Mohammad K, NourBalla AA: An accelerated nutrition transition in Iran. J Public Health Nutr 2002; 5:149–156.

3 Djazayery A, Pajooyan J: Food consumption patterns and nutritional problems in the Islamic Republic of Iran. Nutr Health 2000;14: 53–61.
4 MOHME (Ministry of Health and Medical Education)/UNICEF. The Nutritional Status of Children. October-November 1998. Tehran, MOHME/UNICEF, 1999.
5 Ghassemi H: Report on the 'National Food and Nutrition Security – A Planning and Administration Model' Project. Tehran, National Nutrition and Food Technology Research Institute/Plan and Budget Organization, 1999.
6 FAO/WHO: Diet, Nutrition and the Prevention of Chronic Diseases. Report of a Joint WHO/FAO Expert Consultation. WHO Tech Rep Ser 916. Geneva, WHO, 2003.
7 USDA, Center for Nutrition Policy and Promotion: The Food Guide Pyramid. Home Garden Bull 1996:252.
8 Kimiagar M et al: A comprehensive study on food intake and nutrition in Iran (in Farsi). National Nutrition and Food Technology Research Institute/Plan and Budget Organization, 1999.
9 Painter J, Rah JH, Lee YK: Comparison of international food guide pictorial presentations. J Am Diet Assoc 2002;102:483–489.
10 Orbeta SS: The Filipino Pyramid Food Guide. Nutr Today 1998;33:210–216.
11 Korean Nutrition Society: Recommended Dietary Allowances for Koreans, 7th Revision. Seoul, Jung-Ang Publishing, 2000.
12 The Chinese Nutrition Society: Dietary Guidelines and the Food Guide Pagoda. J Am Diet Assoc 2000;100:886–887.
13 Graca P: Dietary guidelines and food nutrient intakes in Portugal. Br J Nutr 1999; 81(suppl 2):S99–S103.
14 Hunt P, Rayner M, Gatenby S: A national food guide for the UK background and development. J Hum Nutr Diet 1995;8:315–322.
15 National Nutrition and Food Technology Research Institute: How to Eat to Be Healthy. Tehran, Ministry of Health and Medical education/National Nutrition and Food Technology Research Institute/UNICEF, 2001.
16 Canada's Food Guide to Healthy Eating. Ottawa, Health Canada, 1992. H39-252/1992E. Ministry of Public Works and Government Services Canada, 1997. ISBN 0-662-19648-1.

Annals of
Nutrition &
Metabolism

Ann Nutr Metab 2007;51(suppl 2):36–43
DOI: 10.1159/000103566

Published online: July 30, 2007

Visualization of Food-Based Dietary Guidelines – Examples

Juergen S. Koenig

Department of Nutritional Sciences, University of Vienna, Vienna, Austria

Key Words
Visualization · Food-based dietary guidelines · International guidelines · Key concepts · Translation · Nutrient-based dietary guidelines

Abstract
Food-based dietary guidelines (FBDGs) are the primary instrument for the translation of nutrient-based guidelines into practice for the general population. The key concepts for the development of FBDGs as outlined by the report of a joint FAO/WHO consultation in 1996 are to address dietary patterns, to be practical, to be comprehensible, and to be culturally acceptable. These key concepts need to be considered particularly when planning to visualize FBDGs, since the visualization is the major mode of information transfer and the first – and quite frequently the only – contact with the guidelines. For the visualization of dietary patterns it is recommended that total diets rather than nutrients or individual foods should be addressed, although there are some examples which were successful in addressing specific foods and food groups. The design of FBDGs should be flexible for use by people with different lifestyles as well as people of different ages and with different physiological conditions. However, the visualization of FBDGs is a marketing instrument, which differs in no way from regular and commercial marketing instruments. To be successful, these instruments need to be specific for the target population. With increasing power, influence, and coverage of computer-based instruments, the individualization of FBDGs has been facilitated. New strategies have therefore been developed to provide guidelines tailored to individual needs and preferences, as can be seen in the concept of the new food guide pyramids by the US Department of Agriculture. Whether these guidelines are successful needs to be awaited, their major problem being initiative required from the consumer rather than delivering the information appropriately and conveniently to the target population.

Introduction

In 1992, the International Conference on Nutrition in Rome developed the Plan of Action of Nutrition [1], which emphasized the promotion of appropriate diets and healthy lifestyles as one bullet point within the nine action-oriented strategies of this plan. This strategy calls on governments 'on the basis of energy and nutrient recommendations to provide advice to the public by disseminating, through the use of mass media and other appropriate means, qualitative and/or quantitative dietary guidelines relevant for different age groups and lifestyles and appropriate for the country's population' [2]. Facing the differences in the health and nutrition situation in various countries worldwide, it is obvious that those guidelines may diverge considerably from region to region. This paper attempts to provide a short introduction into the concepts of visualization of food-based dietary guidelines (FBDGs) for selected countries and regions.

FBDGs are the primary instrument for the translation of nutrient-based guidelines into practice for the general

KARGER

Fax +41 61 306 12 34
E-Mail karger@karger.ch
www.karger.com

0250–6807/07/0518–0036$23.50/0

Accessible online at:
www.karger.com/anm

Juergen S. Koenig
Department of Nutritional Sciences, University of Vienna
Althanstrasse 14, AT–1090 Vienna (Austria)
Tel. +43 1 4277 54991, Fax +43 1 4277 54999
E-Mail juergen.koenig@univie.ac.at

population. The key concepts for the development of FBDGs as outlined by the report of a joint FAO/WHO consultation in 1996 are to address dietary patterns, to be practical, to be comprehensible, and to be culturally acceptable (table 1) [3]. These key concepts need to be considered particularly when planning to visualize FBDGs, since the visualization is the major mode of information transfer and the first – and quite frequently the only – contact with the guidelines.

For the visualization of dietary patterns it is recommended that total diets rather than nutrients or individual foods should be addressed, although there are some examples which were successful in addressing specific foods and food groups. The 5-a-day program in its national variations is a prominent example for not addressing total diets, but still having had some impact on nutrition behavior and awareness of the general public. It is obvious and within the nature of FBDGs that they need to reflect food patterns rather than numeric nutrient goals. This key concept however is being missed frequently. It is also important not to polarize the target population by visualizing specific diets and also to emphasize that various dietary patterns can be consistent with good health.

In respect to practicability, the recommended foods or food groups should be affordable, widely available and accessible to most people. The recommendation of increased marine fish consumption is idle for populations living far from fishing grounds and lack of logistic infrastructure. Additionally, the recognition of social, economic, agricultural, and environmental conditions affecting foods and eating patterns need to be considered. The design of FBDGs should be flexible for use by people with different lifestyles as well as people of different ages and with different physiological conditions. However, the visualization of FBDGs is a marketing instrument, which differs in no way from regular and commercial marketing instruments. To be successful, these instruments need to be specific for the target population.

Comprehensibility, immediately after first visual contact, is mandatory for visualized FBDGs, in particular to be recognized in the context of other commercials in an information overloaded society. It is obligatory that the guidelines are easily understood by the general public and the target population, respectively, and testing is essential before dissemination.

The major focus in the development of FBDGs is on the cultural acceptability of the recommendations, and this must be reflected in their visualization. The choice of foods and colors in illustrations should be culturally appropriate and sensitive to religious and other cultural considerations, especially those of minority groups. Radical changes in current dietary practices should be avoided and both wording and presentation have to be in the appropriate dialects and languages. It is a psychological standard to be positive and to encourage indulgence of appropriate diets.

Table 1. Key principles for developing FBDGs

Dietary patterns
• Total diet, rather than nutrients or individual foods, should be addressed
• Dietary guidelines need to reflect food patterns rather than numerical nutrient goals
• Various dietary patterns can be consistent with good health
Practicality
• The recommended foods or food groups should be affordable, widely available and accessible to most people
• FBDGs should recognize the social, economic, agricultural and environmental conditions affecting foods and eating patterns
• FBDGs should be flexible for use by people with different lifestyles as well as people of different ages and different physiological conditions
Comprehensibility
• FBDGs should be easily understood by the general public, taking into account levels of literacy. The terminology used must be simple and must refer to foods, rather than nutrients, wherever possible
• Food groups that make sense to the public should be chosen
• Visual presentation must be easily understood
• Testing of the FBDGs is essential before dissemination
Cultural acceptability
• The choice of foods and colors used in illustrations should be culturally appropriate
• The guidelines should be sensitive to religious and other cultural considerations, especially those of minority groups
• FBDGs should not recommend radical changes in current dietary practices
• Presentation should use appropriate language or dialect
• FBDGs should be positive and should encourage enjoyment of appropriate diets

The Food Guide Pyramid

Attempts to translate nutrient-based recommendations into visually comprehensible models of FBDGs have been made by various countries, frequently by adopting existing models to national and regional distinctions. The effort made by the United States Depart-

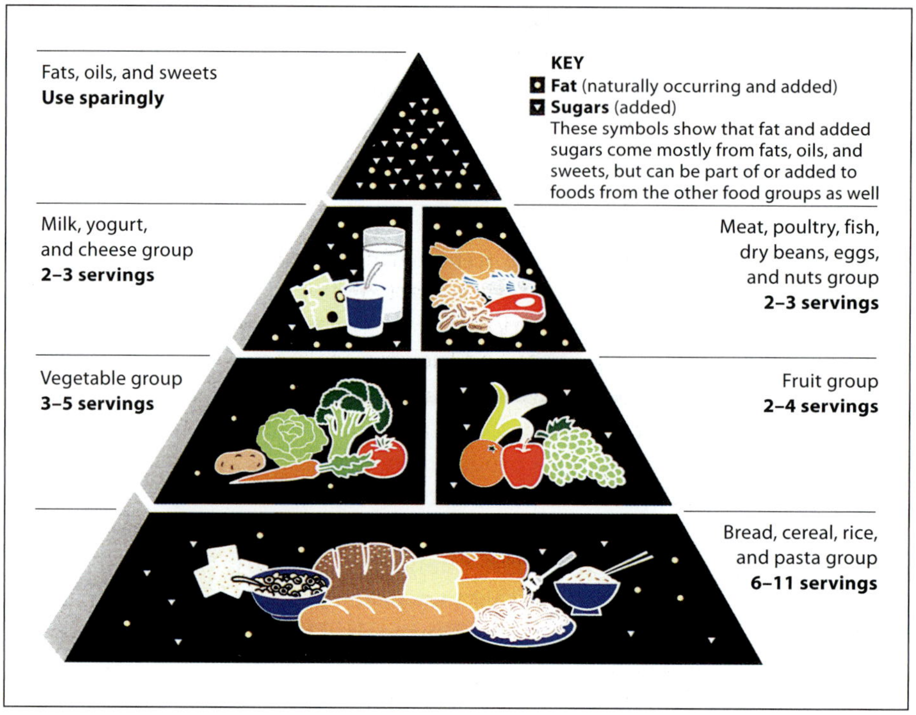

Fig. 1. The Food Guide Pyramid from the USDA from 1992 [4].

ment of Agriculture (USDA) for the development of the Food Guide Pyramid (fig. 1) [4] in the 1990s and the efforts put into harmonization of these guidelines across the North American countries placed the Food Guide Pyramid at a strong position as a role model for other countries worldwide. A WHO evaluation of FBDGs from 30 countries demonstrated that 16 of the 35 guidelines use a food pyramid [5]. The beginning of the Food Guide Pyramid's doom was the adoption by the food industry as an instrument of product-specific marketing strategies, thus leading to a tremendous loss of credibility in the public and subsequently to the development of rivalry within the sector of FBDGs. The strongest impact at the end came from the competing food pyramid developed by Walter Willett in his book 'Eat, Drink and Be Healthy' in 2001 (fig. 2) [6], in which he strongly opposed the well-established USDA Food Guide Pyramid mainly for its emphasis on carbohydrates and alleged demonization of fats.

Unfortunately, the controversial discussion of the two guidelines in the public counteracted the yield of the USDA Food Guide Pyramid gained by the unambiguous support throughout wide parts of the nutrition community and its positive impact was lost within weeks, demonstrating the susceptibility of the nutrition information market. Finally, the debate ended in the development of

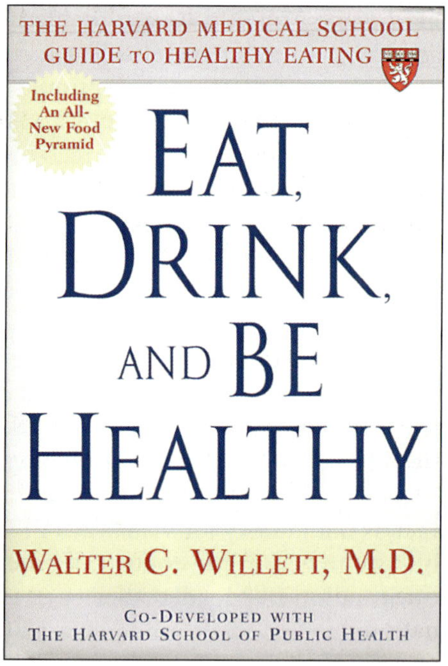

Fig. 2. 2001 book cover of 'Eat, Drink and Be Healthy' [6].

a modified concept of the USDA Food Guide Pyramid, the new MyPyramid.gov campaign (fig. 3) [7].

Some controversy still exists in the discussion of the appropriate design of the principle set-up of the food guide pyramid and its derivates in the various countries. Most pyramids show those foods at the base of the pyramid which predominantly supply carbohydrates (cereals, bread, rice, pasta, and the like), thus indicating this food group to be the quantitatively largest part of the diet. Fruit and vegetable recommendations are the second level of the food guide pyramid based on the amount of recommended servings. From a pure quantitative point of view, however, fruit and vegetables combined should form a larger part of the diet compared to the carbohydrate group. This has caused some confusion in the general public, again decreasing credibility of the entire concept. The situation even declines when including fluid intake into the guidelines, now resulting in fluids at the base of the pyramid. Some attempts have been made to work around these controversies, again leading to different models which in some cases have been disseminated concurrently for identical target populations.

With increasing power, influence, and coverage of computer-based instruments, the individualization of FBDGs has been facilitated. New strategies have therefore been developed to provide guidelines tailored to individual needs and preferences, as can be seen in the concept of the new food guide pyramids by the USDA. Whether these guidelines are successful needs to be awaited; their major problem being initiative required from the consumer rather than delivering the information appropriately and conveniently to the target population.

National Concepts

In view of the cultural aspects of FBDGs, it is obvious that different countries and regions have developed diverging models for the dissemination and visualization of nation- or region-specific guidelines. It is unfortunately impossible within the scope of this paper to include images of the various examples worldwide. Many of the national FBDGs have been collected by the FAO and can be assessed through their website [8].

Based on a WHO report on the global status of FB DGs, 75 countries of 126 reviewed provide a final version of guidelines for their population, 12 countries in the process of preparing guidelines, and 39 have provided no information [5]. The analysis of 64 sets of guidelines from 37 countries showed that in most cases a target group is either not specified or the general population is addressed. In the same sample of guidelines, almost 80% include messages on fruit and vegetables, some 50% disseminate messages on milk and dairy products and about 30% recommendations for fish consumption. In many cases, selected nutrients are the focus of the messages, salt, fats, and sugar being the predominant topics addressed. Frequently, FBDGs also convey messages on issues not directly related to diet. Physical activity, body weight, food safety and hygiene, breastfeeding, and financial aspects are addressed in many of the sets of guidelines. Some 90% of the guidelines reviewed also include at least one social or behavioral message, e.g. enjoy your meals, make mealtime a social time, and eat a variety of foods every day. The evaluation of the messages on fruit and vegetable intake showed that 43 of the 50 messages were different to some extent, only seven delivered the same message (eat plenty of fruit and vegetables).

Fig. 3. MyPyramid.gov [7].

The approaches followed by China, Iran, Austria, Switzerland, and Germany are discussed in detail elsewhere in this issue. Therefore, some selected examples from other countries are discussed in the following.

A very similar approach to the food guide pyramid of the United States FDA has been followed by the Working Group on Food-Bases Dietary Guidelines for Thai People in the Department of Health of the Ministry of Public Health in Thailand [9]. Their concept on 'Healthy Eating for Thais' includes a Nutrition Flag, which is basically identical to the Food Guide Pyramid but has been turned upside down. This approach is appealing in respect to the typical reading direction from top to bottom in contrast to the food pyramids, and thus directing the reader's emphasis to the quantitatively most important food groups on first sight of the guidelines. The Thai Nutrition Flag depicts examples for the recommended food groups by selecting typical foods consumed by the Thai population which is in accordance with the FAO/WHO key concepts for the development of FBDGs. The bread, cereal, rice and pasta group from the US food guide pyramid has been converted into a group of rice-starchy food and the servings have been adopted by using rice-serving spoons as measuring unit. Again based on typical consumption patterns, the milk, yoghurt and cheese group has been reduced to a milk only group, since other milk products are of less importance in the Thai diet.

The visual concept of the Thai nutrition flag is accompanied by worded guidelines with nine rules as listed in table 2. Again in analogy to the USDA Food Guide Pyramid, the Thai FBDGs provide more detailed recommendations by visualizing examples for the recommended food groups within the Nutrition Flag and giving assistance for the conversion of household measures into the recommended servings. One plate of rice is consequently equal to three rice-serving spoons of rice, one bowl of noodles equals two rice-serving spoons, and one slice of bread, one piece of fermented noodles and one ear of corn, respectively, each represent one rice-serving spoon of those foods.

The Thailand experience in developing their FBDGs showed the importance of many of the issues addressed by FAO/WHO in their recommendations [10]. An expert working group had been established for the development of the guidelines, which collected data on food consumption and public health-related issues prior to the development of the Thai guidelines, and data from other countries had additionally been reviewed by the working group. Similar to other countries, the working group recognized that, although appropriate data is limited, there still is no justification to abandon the process of developing national FBDGs. Institutional commitment had been identified a key issue for the process and outcome of the guidelines. Support and endorsement came from high-level governmental participation in the process of guideline development and led to the required media attention. However, some concern was expressed about the maintenance of the integrity of the messages, in particular in view of food industry using subsets of the guidelines to promote their own products. Inconsistencies had been found between the Thai Nutrition Flag and guidelines developed by private, non-governmental organizations.

Japan issued their FBDGs in a similar format as the Thai Nutrition Flag, and disseminated also a reversed food pyramid in form of a spinning top which is claimed to be a well-known traditional Japanese toy [11]. This has been designed to symbolize physical activity as part of the guidelines both by the dynamic motion of the spinning top itself and a pictogram of a running person on the top of the spinner. The spinning top as a symbol of balance is appealing as it may remind the target population of the importance to maintain an optimal balance in their diets. The handle of the spinning top represents fluid intake by water and tea, thus adding recommendations on fluid intake into the concept but disregarding the quantitative proportionality. Snacks, confection and beverages, which are considered non-essential treats are represented by a string attached to the side of the spinning top, symbolizing the momentum the string transfers to the spinner and the supposed momentum of these foods transferred to the Japanese diet.

Table 2. FBDGs for Thailand [9]

Rule 1	Eat a variety of foods from each of the 5 food groups and maintain proper weight
Rule 2	Eat an adequate amount of rice or alternative carbohydrate sources
Rule 3	Eat plenty of vegetables and fruits regularly
Rule 4	Eat fish, lean meats, eggs, legumes and pulses regularly
Rule 5	Drink milk in appropriate quality and quantity for one's age
Rule 6	Eat a diet containing appropriate amounts of fat
Rule 7	Avoid sweet and salty foods
Rule 8	Eat clean and safe foods
Rule 9	Avoid or reduce consumption of alcoholic beverages

Interestingly, the spinning top from Japan shows fruit at the tip of the spinning top together with milk and milk products in two separate groups. Although two servings are recommended to be consumed daily for both groups, the position at the tip implies that these food groups are of less importance than the other groups at the upper part of the spinning top, like grain dishes (5–7 servings per day) at the top and vegetable dishes (5–6 servings per day) at the second level.

The Japanese Food Guide Spinning Top is embedded into the broader concept of food education law 'Shokuiku' by the Japanese Ministry of Agriculture, Forestry and Fisheries in 2005 [11]. The basic concepts of Shokuiku are listed in table 3.

The spinning top is the visualized set of dietary guidelines developed within the same frame of food education as the spinning top itself. These guidelines are similar to dietary guidelines from other countries, demonstrating the evidence of comparable nutrition problems in the Jap-

Fig. 4. French Nutrition Guide Health Comes From Nutrition [12].

Table 3. Basic concepts of the Japanese food education law 'Shokuiku'

1	Shokuiku should have the purpose of promoting people's health in body and in mind, as well as enriching human lives
2	Care should be taken to induce people to develop greater appreciation for and understanding of their diets, including the various roles played by the natural environment and the people who produce, transport and prepare food
3	Nationwide, voluntary movements for promotion of Shokuiku should be developed
4	Especially parents, educators and day-care providers should actively promote Shokuiku among children
5	Understanding of dietary issues should be reinforced by taking advantage of all available opportunities, such as at home, in schools and in the community or everywhere, to offer a variety of food-related experiences and activities
6	Awareness and appreciation of traditional Japanese food culture as well as food supply/demand situations should be promoted, and opportunities of interaction between food producers and consumers should be created, in order to revitalize rural farming and fishing regions, and to boost food self-sufficiency in Japan
7	Appropriate information on food safety should be offered, in order to help people to practice proper diets

anese population. The dietary guidelines for Japanese are:

- Enjoy your meals
- Establish a health rhythm by keeping regular hours for meals
- Eat well-balanced meals with staple food, as well as main and side dishes
- Eat enough grains such as rice and other cereals
- Combine vegetables, fruits, milk products, beans and fish in your diet
- Avoid too much salt and fat
- Learn your healthy body weight and balance the calories you eat with physical activity
- Take advantage of your dietary culture and local food products while incorporating new and different dishes
- Reduce leftovers and waste through proper cooking and storage methods
- Assess your daily eating.

A different approach to FBDGs has been followed by France to some extent. The French Ministry of Health and Solidarity (Ministère de la Santé et des Solidarités) issued a comprehensive 127-page booklet entitled 'Health Comes From Nutrition' (La Santé vient en mangeant) [12]. The guide attempts to cover the complete range of nutritional aspects in the different physiological, social, and economical situation of the French population. Although no small set of visualized guidelines has been developed, the symbolism of the French guide does include a clear line throughout the brochure by starting each chapter with recommended foods laid out in the style of the Italian painter Archimboldo and his portrait heads (fig. 4).

The French FBDGs are composed of 25 chapters with the headlines as listed in table 4. In its introductory remarks, the guide gives as an explanation for the reason of a food guide that nutrition is one of the great pleasures of life, leading from the basic pleasure of simply not being hungry, to the higher pleasures of culinary indulgence, the joy in sharing meals with family and friends, or the gratification of not having to worry about nutrition. This is an interesting and very appealing way of putting the enjoyment of nutrition on top of the recommendations and very much represents the French attitude towards nutrition. Nonetheless, simply the length of the guide may be inappropriate for the general population and this may mislead to extract parts of the guide for specific groups of the population or specific physiological or social situations. Thus, important parts of the information provided by the guide may be missed from those concerned by various nutritional issues.

Canada developed their 'Food Guide to Healthy Eating For People Four Years and Over' in the form of a rainbow-

Table 4. Chapters and headlines covered in the French FBDGs 'La santé vient en mangeant'

Chapter	French headline	English translation
1	Je veux: manger, protéger ma santé ... et me faire plaisir!	I want to: eat, protect my health ... and have fun!
2	La bouffe, je m'en fous	Nutrition, I don't care
3	Je prépare à manger pour ma famille	I prepare food for my family
4	J'ai du mal à joindre les deux bouts	I have problems to make ends meet
5	Je mange un sandwich tous les midis	I only have a sandwich for lunch
6	Je mange souvent au restaurant	I often eat at restaurants
7	Je mange souvent au fast-food	I often eat at fast food restaurants
8	Je saute des repas, je n'ai pas le temps de manger	I skip meals, when I don't have time
9	Je ne cuisine pas	I don't cook
10	J'ai tendance à grignoter entre les repas	I tend to nibble before the meals
11	J'ai toujours faim	I'm always hungry
12	Je n'aime pas certains aliments	I don't like some foods
13	Je reçois souvent, j'aime faire la fête	I like to party
14	Je ne mange que des plats cuisinés du commerce	I only eat convenience foods
15	Je me restreins pour ne pas grossir, je suis au régime	I limit myself on fattening stuff, I am on diet
16	Je mange souvent du fromage et/ou de la charcuterie	I often eat cheese and/or sausages
17	Je raffole des aliments salés de type «apéritif»	I fancy salty snacks as an entree
18	J'aime boire	I like to drink
19	Je mange souvent des plats exotiques ou d'origine étrangère	I often eat exotic or foreign food
20	Je ne mange que du «bio»	I only eat organic foods
21	Je suis végétarien	I am vegetarian
22	Je ne mange pas certains aliments pour des raisons religieuses	I don't eat some foods for religious reasons
23	J'ai peur de la vache folle, des OGM, de la dioxine ...	I am afraid of mad cow disease, genetically modified organisms, dioxin, ...
24	Je suis plutôt sédentaire, je ne fais pas de sport ...	I am rather settled, I don't do any sports
25	Je suis plutôt sportif	I am rather sporty

like quarter circle [13]. The red-colored inner section of the rainbow represents meat and alternatives, the adjacent blue section stands for milk products, the third green section for vegetables and fruits, and the outer section in yellow lists grain products as largest part of the rainbow. Together with pictures of the different foods of the four sections, the Canadian food guide states to enjoy a variety of foods from each group every day and to choose low-fat foods more often. Like other food guides, the Canadian food guide rainbow is supplemented by detailed recommendations on the amounts of servings and typical serving sizes for the four food groups. Grain products are recommended to be consumed from 5–12 servings per day, while 1 serving may be e.g. 1 slice of bread, 30 g of cold cereals, 175 ml of hot cereals, etc. 5–10 servings of vegetables and fruits are suggested and to choose dark green and orange vegetables and orange fruit more often. In contrast to other examples of national FBDGs, the Canadian food-guide rainbow distinguishes its guideline on milk products by age groups: children of 4–9 years are recommended to daily consume 2–3 servings, youth 10–16 years and pregnant women should have 3–4 servings, and adults 2–4 servings. It is also recommended to choose low-fat milk products more often. Similarly, low-fat meat and alternatives are recommended in form of leaner meats, poultry and fish, but also dried peas, beans and lentils.

In 2004, an extended version of the food-guide pyramid has been issued in form of the Canadian Guide to Healthy Eating and Physical Activity [14]. The quarter circle rainbow has been complemented to a half circle rainbow by adding recommendations on physical activity, advising to reduce sitting for long periods, to increase strength activities, to increase flexibility activities, and to increase endurance activities. Consequently, the Canadian Guide provides the most detailed and comprehensive recommendation on physical activity and puts this equal to the recommendations on nutrition by using the same area and format.

The FBDGs developed for the UK in form of the food plate has also been used as a role model for other countries (e.g. Australia) and can also be seen as the model used for the development of the German food circle. The original guideline shows the recommended food groups fruit and vegetables, bread cereals and potatoes, milk and dairy food, meat, fish and alternatives, and fatty and sugary foods according to their quantity recommended [15]. This visualization of the UK FBDGs comes together with eight worded guidelines, which are similar to those of other countries. These guidelines are:

- Base your meals on starchy foods
- Eat lots of fruit and vegetables
- Eat more fish
- Cut down on saturated fat and sugar
- Try to eat less salt – no more than 6 g a day
- Get active and try to be a healthy weight
- Drink plenty of water
- Don't skip breakfast.

The UK guidelines are a good example of the transferability of the concept to other countries and societies which are similar in cultural background and have similar nutrition issues to address. However, it is obvious that the process of adopting guidelines needs also to consider the particularities of the target population to gain the required acceptance in populations of different nutritional background.

Concluding Remarks

FBDGs are – if done properly – an elaborative task and require substantial amount of developmental work for many expert groups from various disciplines and responsibilities. They have to be regarded in the same way as any other marketing instrument to be successful. Enormous resources are therefore required not only for the development but also for the dissemination of the guidelines on a regular basis and with appropriate mass media coverage. In the presence of a wide range of food and nutrition information, whether brand- and product-specific or independent from industry interests, FBDGs need to stand out from the information overload in the nutrition sector.

It is mandatory in view of our globalized society that FBDGs are harmonized internationally, to avoid the intrusion of diverging guidelines from different countries. Obviously, the national or regional harmonization is crucial and should include information issued from governmental bodies, non-governmental organizations, and the private sector. FBDGs require constant development and adoption, to respond to a changing society, but this adoption has to proceed gently and wisely.

It is unfortunate that the nutrition community is inconsistent in its public appearance. Therefore, close networking involving all groups dealing with nutritional issues is vital to plan for clear and unbiased messages for the general public. Nutritional science, as any other natural science, is in a process of constant development and new findings may lead to changing evidence. However, this is a subtle process and sudden paradigm changes are actually rare to occur. Although it is difficult to disseminate information on a scientifically appropriate level with its amount of details required for scientific accuracy, it is important not to polarize the guidelines for the sake of comprehensibility. At the present state, it is obvious that FBDGs require a lot more resources to be successful and effective in improving nutrition of the target populations.

References

1 Food and Agriculture Organisation, World Health Organisation: International Conference on Nutrition. World Declaration on Nutrition and Plan of Action for Nutrition, Rome 1992.
2 World Health Organisation: Preparation and Use of Food-Based Dietary Guidelines. Report of a Joint FAO/WHO Consultation. WHO/NUT/96.6. Geneva, WHO, 1996.
3 Clay WD: Preparation and Use of Food-Based Dietary Guidelines. FNA/ANA 1997; 19:42–47.
4 United States Department of Agriculture: The Food Guide Pyramid. Washington, USDA, 1992.
5 Nishida C: WHO's Work on Development and Implementation of Food-Based Dietary Guidelines. EFSA Colloquium 5: Development of Food-Based Dietary Guidelines, Parma 2006.
6 Willett WC: Eat, Drink and Be Healthy. New York, Free Press, 2001.
7 United States Department of Agriculture: MyPyramid.gov – Steps to a Healthier You. http://www.mypyramid.gov (accessed 28.10.2006).
8 Food and Agriculture Organisation: National Dietary Guidelines. http://www.fao.org/ag/agn/nutrition/education_guidelines_en.stm (accessed 28.10.2006).
9 Ministry of Public Health: Manual Nutrition Flag. Healthy Eating for Thais. Working Group on Food-Based Dietary Guidelines For Thai People. Bangkok, Nutrition Division/Department of Health, 2001.
10 Schneeman BO: Preparation and Use of Food-Based Dietary Guidelines: Lessons from Thailand and the Philippines. FNA/ANA 2001;28:55–64.
11 Ministry of Agriculture, Forestry and Fisheries of Japan: What Is 'Shokuiku (Food Education)'? Tokyo, MAFFJ, 2005.
12 Ministry of Health and Solidarity of France: La Santé Vient en Mangeant. Paris, MHSF, 2002.
13 Ministry of Public Works and Government Services Canada: Canada's Food Guide to Healthy Eating for People Four Years and Over. Ottawa, MPWGSC, 1997.
14 Government of Canada: Canadian Guide to Healthy Eating and Physical Activity. Ottawa, Government of Canada, 2004.
15 Department of Health of the United Kingdom: Food and Health Action Plan. London, DHUK, 2003.

Annals of
Nutrition & Metabolism

Ann Nutr Metab 2007;51(suppl 2):44–49
DOI: 10.1159/000103567

Published online: July 30, 2007

Nutrient Profiles – A Suitable Means of Nutrition Information

Helmut F. Erbersdobler

Institute of Nutrition and Food Science, University of Kiel, Kiel, Germany

Key Words
Health claims · Thresholds · Scoring system · Evaluation 'across the board' · Food categories · Consumer protection · Disqualifying nutrients

Abstract
For several years, nutrient profiling has been discussed as a regulatory instrument for health claims made on foods. The aims of the European Commission are mainly to protect consumers from misleading and even dangerous information and to ban advertisements for foods and menus for children if they are not healthy. Generally, nutrient profiling will be established on the basis of contents of different nutrients and substances like fat, saturated fat, trans fatty acids, salt/sodium and sugars 'whose excessive intakes in the overall diet are not recommended and those such as poly- and monounsaturated fats, available carbohydrates other than sugars, vitamins, minerals, protein and fiber'. In the present paper, 4 out of >20 proposals will be discussed: the US requirement for health claims (FDA Code), the 'a little a lot' system from the UK, the 'tripartite classification' model of the Netherlands and the FSA scoring system from the UK. At present, several details remain uncertain, e.g. the reference basis (100 g, serving size or 100 kcal, respectively), moreover the use of thresholds or a scoring system. The evaluation may be 'across the board' versus in categories, respectively. In all cases, exceptions will be necessary.

Introduction

The term *nutrient profiling* is being intensively discussed since the European Commission proposed it as a regulatory instrument for health claims made on foods. According to the new proposal, nutrient- and health-related claims should only be allowed if the items in question fulfill the scores of specially defined nutrient profiles. The aims of the Council of the European Union and the European Parliament are (1) to protect consumers from misleading and even dangerous information; (2) to ban advertisements for foods and menus for children if they are not healthy in the sense of a good nutrient profile, and (3) 'creating equal conditions of competition for the food industry' [1].

Generally there are several possibilities to characterize foods in order to make food-based recommendations:

- General statements or warnings (e.g. 'added sugar', 'rich in saturated fats')
- Position in the Food Pyramid
- Simple dietary guidelines (ex. 10 Regeln der DGE [2], 'Nutrition and Your Health', USDA 2005 [3])
- A traffic light system – red, yellow, green stickers
- Use of nutrient profiles or a similar parameter.

Nutrient profiles were well known in some special fields of nutrition like protein evaluation or fatty acid profiling. However, for a judgment of the whole food quality the total number of nutrients would be too much. For the sake of clarity the number of nutrients included in the evaluation should be limited to the most important

KARGER
Fax +41 61 306 12 34
E-Mail karger@karger.ch
www.karger.com

0250–6807/07/0518–0044$23.50/0

Accessible online at:
www.karger.com/anm

Helmut F. Erbersdobler
Institut für Humanernährung und Lebensmittelkunde der
Christian-Albrechts-Universität zu Kiel, Duesternbrooker Weg 17
DE–24105 Kiel (Germany), Tel. +49 431 880 5680, Fax +49 431 880 5679
E-Mail erbersdobler@nutrfoodsc.uni-kiel.de

ones. Additionally, the data should be related to reference values in order to avoid too many dimensions (from μg to g). Generally, the European Commission has proposed to establish the nutrient profiles on the basis of 'the content of different nutrients and substances with physiological effect, in particular those such as fat, saturated fat, trans fatty acids, salt/sodium and sugars whose excessive intakes in the overall diet are not recommended and those such as poly- and monounsaturated fats, available carbohydrates other than sugars, vitamins, minerals, protein and fiber' [1]. Among these positive nutrients, dietary fiber, folic acid, n–3 fatty acids, calcium and iron are discussed. In foods for elderly people, vitamins D and B_{12} would also be of interest. According to the German Nutritional Report 2004 [4], German adult men have low intakes of folic acid and dietary fiber, adult women additionally partially calcium and iron (low values for vitamin D are not considered because of the unknown contribution of vitamin D synthesis after UV radiation). Iodine is only limited in the Central European region if no iodinated salt is used.

At the present time the following questions are being discussed concerning several proposals:

- How many nutrients should be selected and especially which ones – only 'negative' nutrients, or 'negative' plus 'positive' nutrients?
- Should differences between age groups or gender variations be considered?
- Which reference values should be taken, national data or common values established by the European Union or WHO/FAO?
- Reference basis – 100 g; serving size; 100 kcal?
- Use of thresholds or scores?
- Category-specific or an evaluation 'across the board'?

The most important points seem to be the last three ones. At present it appears that a threshold solution will be favored which leads to a more category-specific view. The present proposals favor the basis of 100 g for defining nutritional profiles. However, the possible use of all three dimensions – 100 g, 100 kJ/kcal, serving sizes – are still under discussion. Until now, 22 different proposals have been discussed, of which 3 use only negative criteria but 18 negative and positive aspects of nutrient concentrations. Six favor food categories and 15 an evaluation across the board respectively. Seventeen voted for a threshold and 4 for a scoring system. Eight use a 100-gram basis, 11 use serving sizes and 2 mixtures of 100 g + 100 kcal or 100 kcal + serving size, respectively. Some of the questions will be discussed below.

Table 1. US requirement for health claims. Thresholds of different nutrients calculated on a basis of serving sizes (FDA Code of Federal Regulations [5])

Maximum (disqualifying)		Minimum (qualifying)	
Total fat	13 g	Vitamin A	500 IU
Saturated fatty acids	4 g	Vitamin C	6 mg
Cholesterol	60 mg	Iron	1.8 mg
Sodium	480 mg	Calcium	100 mg
		Protein	5 g
		Dietary fiber	2.5 g

Description of Four Profiling Models Using Different Criteria

In the USA, the FDA has specified criteria for a profiling model (FDA Code of Federal Regulations [5]). This system uses thresholds for disqualifying and qualifying nutrients per serving. The criteria are judged across the board; however, disparities between the food categories are considered by accounting to the serving sizes. Moreover, there are a lot of exemptions which are not discussed in detail here. The disqualifying and qualifying nutrients and their maxima and minima are shown in table 1. The values correspond to thresholds for the disqualifying nutrients of <20% of the daily recommended value and for the qualifying nutrients of >10%, respectively (calculated for a 2,000-kcal diet [6]).

As an example for the FDA system, 30 g of cornflakes were evaluated as shown in figure 1 [data from 7]. This item contains a lot of sodium which would be a problem if 100 g would be consumed (and calculated). However, by using the serving size of 30 g for that item the threshold for sodium is not exceeded. As for allowance to carry a health claim the food has to satisfy at least one of the qualifying minima, and this is the fact for iron. As a result, a claim is possible.

The 'a little a lot' system from the UK is a similar proposal using also thresholds across the board but on a 100-gram basis [8]. It uses one-fifth of the relative Guideline Daily Amount (GDA) as threshold for 'a lot' (calculated for a 2,245-kcal diet), which is >10 g of non-milk extrinsic sugar, >20 g of fat, >5 g saturated of fatty acids, >0.5 g of sodium and >3 g of NSP fiber, respectively. This is quite similar to the data of the above-mentioned US system. On the other side, the threshold values for the 'a little' criterion are much lower with one-thirtieth of the relative GDA, which is <2 g for non-milk extrinsic sugar,

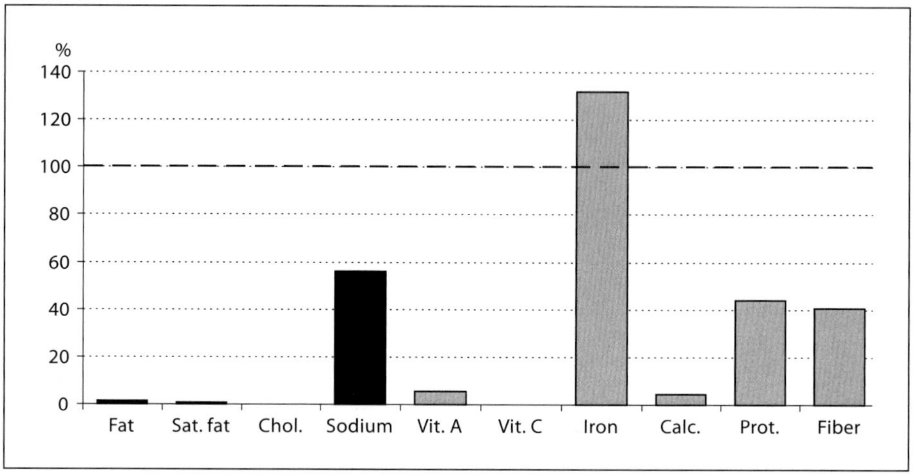

Fig. 1. An example (30 g cornflakes) for the US requirement for health claims (FDA Code); result: claim possible!

Table 2. Criteria of the UK FSA scoring system. Points depending on quantity of nutrients per 100 g

	Points										
	0	1	2	3	4	5	6	7	8	9	10
A. Negative nutrients											
Energy, kJ	≤335	>335	>670	>1,005	>1,340	>1,675	>2,010	>2,345	>2,680	>3,015	>3,350
Saturated fat, g	≤1	>1	>2	>3	>4	>5	>6	>7	>8	>9	>10
Total sugar, g	≤4.5	>4.5	>9	>13.5	>18	>22.5	>27	>31	>36	>40	>45
Sodium, mg	≤90	>90	>180	>270	>360	>450	>540	>630	>720	>810	>900
B. Positive criteria											
Fruits, vegetables and nuts, %	≤40	>40	>60	–	–	>80					
NSP fiber, g	≤0.7	>0.7	>1.4	>2.1	>2.8	>3.5					
or AOAC fiber, g	≤0.9	>0.9	>1.9	>2.8	>3.7	>4.7					
Protein, g	≤1.6	>1.6	>3.2	>4.8	>6.4	>8					

<3 g for fat, <1 g for saturates, <0.1 g for sodium, and <0.5 g for fiber, respectively. Since this procedure gives no clear decision for evaluation, it is not demonstrated with a practical example. However, the result of an evaluation of cornflakes with this system would show an exceeding of the threshold 'a lot' for sodium since this system uses 100 g as basis. This could disqualify the item from allowance to carry a claim.

The 'tripartite classification' model of the Netherlands [9, 10] uses negative and positive nutrients in variable selection according to a classification in categories. For discrimination, thresholds are used at a food basis of 100 g. The tripartite classification is: (a) 'preferable', (b) 'middle course', and (c) 'exceptional'. Nutrients and thresholds are different in the food groups (e.g. milk and milk products or cheese or bread, bread substitutes, breakfast cereals, respectively). There are exemptions for food groups with a low nutrient density or high-energy density like snacks, spicy filling, sweets and cream. For these groups only thresholds for the content in saturated fatty acids (low and high) and in fiber (if 'high') are defined. Also for this system there exists no final decision for an application on legislations for claims yet. The primary aim was to inform consumers to make 'healthy' food choices.

Finally, the FSA scoring system for children [11, 12] should be discussed here. This model was first developed for children but can also be used for adults. A preliminary proposal [13] was presented and discussed earlier [14]. The FSA uses a scoring system across the board on a 100-gram basis. Up to 10 points are given for the 'negative'

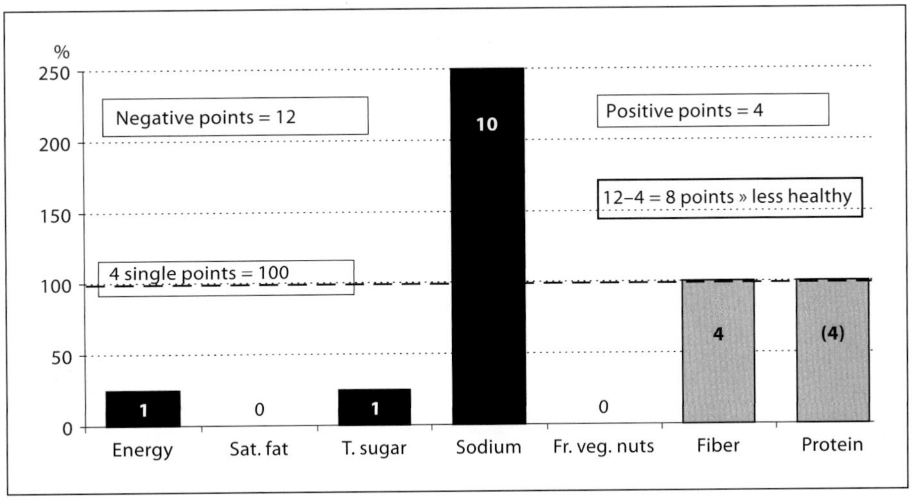

Fig. 2. An example (100 g cornflakes) for the UK FSA scoring system; result: less healthy!

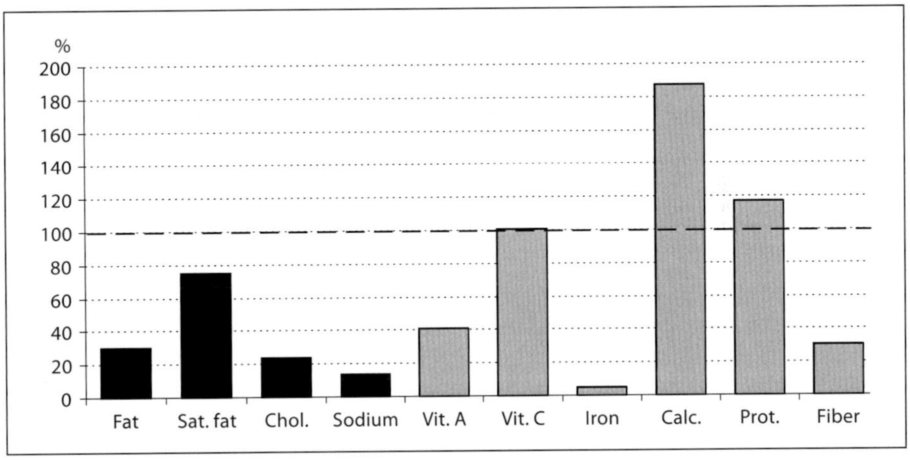

Fig. 3. An example (150 g fruit yoghurt) for the US requirement for health claims (FDA Code); result: claim possible!

nutrients as shown in table 2. Moreover, another 5 points are given for the positive criteria: fruits, vegetables and nuts in %; fiber in g and protein in g as is also shown in table 2. One point corresponds to 3.75% of the UK GDA. The total of points from the positive criteria is subtracted from the total of points from the negative nutrients. However, if the sum of negative points is 11 or higher, then the points from protein cannot be regarded, unless there are 5 points from fruits, vegetables and nuts. The foods are classified as 'less healthy' if the total of points is 4 and more and as 'intermediate' if 1–3 points are reached and 'a healthier choice' where zero or less points are calculated. For drinks, only two classifications are scheduled, 'less healthy' if it has 1 point or more, and 'healthier choice' if it scores zero points or less.

Figure 2 shows again a calculation of the scoring of the cornflakes example. The figure shows that for that item 12 negative points are counted and 8 positive. However, since the negative points sum up to 12 (>10), the positive points of protein cannot be regarded. For this reason the result of subtraction of 4 points from 12 is 8 points and the decision is 'less healthy'. The item cannot bear a claim. On the other hand, the sodium content of the cornflakes appears quite high and some reduction seems to be possible. A reduction of one-third (~300 mg) would lead to only 7–8 points for sodium and to a total of 9–10 negative points. In this case, both positive 4 points for fiber and protein can be used and the result of 9–10 minus 8 would be 1–2 which means a result of 'intermediate'. In this case the cornflakes could bear a claim.

Further Examples

Calculations with the US FDA Code and the UK FSA scoring system for another food item, a yoghurt with fruits (2.6% fat), is shown in figures 3 and 4. The result is that in both cases, this item, 150 g for the US FDA Code and 100 g for the UK FSA scoring system, can bear a claim.

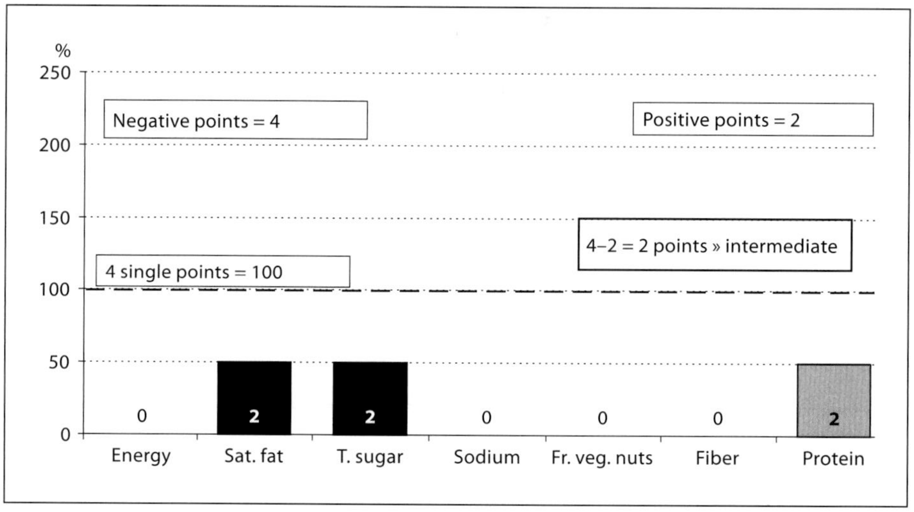

Fig. 4. An example (100 g fruit yoghurt) for the UK FSA scoring system; result: intermediate!

Discussion of the Unsolved Questions

In this paper only a few examples of unsolved questions can be regarded. Far more discussions took place at the ILSI Europe Workshop, April 2006, whose results will be published soon [15].

As we have seen, all systems use different nutrients in more or less different threshold or scoring concentrations. Moreover, some nutrients like dietary fiber and especially the term sugar are used or defined differently. Even the wording for the methods or results is different. Another reason for discrepancy is the use of thresholds versus scoring systems, respectively. In the threshold models foods will be classified as 'less healthy' or 'a healthier choice' depending on whether selected nutrients will be above or below distinct thresholds. Thresholds are relatively easy to understand and to explain to the consumer. On the other hand, they are often oversimplistic and biased since one nutrient can determine the value of a food. In the case of fat, foods with a higher fat content like nuts will be excluded. For this reason more exemptions than in scoring systems will be necessary. Scoring systems use, if they are not too mathematically complicated, a series of different thresholds scoring them with a system of points. A food that would exceed the threshold in the threshold method may be saved in the scoring system if its levels of other nutrients will lower the score. The scoring systems are more flexible, accurate and scientifically based. They can be used for different purposes like comparison of different foods or in nutritional education and for innovations in the food industry. On the other hand, scoring systems are more complex and difficult to understand. Currently the threshold models are favored, but the scoring systems should also be considered and developed.

The evaluation across the board is easier to develop, to control and to change. It allows a better comparison between foods also from different categories. On the other hand, not all critical nutrients fit into all food categories (e.g. fiber to milk products). For this reason more nutrients than in a model with categories would be necessary. Generally the evaluation across the board would need more exemptions.

The evaluation in food categories can use only a few, eventually only one typical criteria. Since certain food categories are higher in certain nutrients than others and vice versa, a specific evaluation would be possible. This would be easier for consumers to understand who would otherwise possibly ask why wholemeal bread may be a healthier choice, although it contains hardly any calcium. Moreover, this would also increase the incentive of the food industry to improve their products. On the other hand, the food categories are also not unique and it would not be easy to define the different groups. Moreover, the expert groups would be fraught with the management of definitions, grouping and allocations of categories and criteria. The US and UK experts seem to vote for an evaluation across the board, while the Netherlands and also German authorities favor models with categories.

Finally, many experts think that only models using 100 g as base should be favored since serving sizes are difficult to define. On the other hand, the systems using service sizes regard the problems of items used in small quantities and avoid by this way several exemptions.

The Aim of Nutrient Profiles

The general aim of the authorities establishing the nutrient profiles is a benefit for consumers by identifying foods and drinks eligible for health claims:

- To lead to a better understanding of food quality and to help consumers make reasonable food choices
- To ban advertisements for foods and menus for children if they are not healthy in the sense of a good nutrition profile
- To make healthy marketing programs directed at children
- To protect consumers from misleading and even dangerous information.

Moreover, the nutrient profiles may also be of benefit for science by evaluating food quality and by using it for food-based dietary guidelines. Well-defined and highly informative nutrient profiles may be useful in nutrient counseling, e.g. for food-based dietary recommendations by supporting visualized information with an easily understandable score, e.g. for explaining the locations of foods in dietary pyramids. However, for this purpose the scoring systems seem to be better suited than the methods using thresholds.

The benefit of nutrient profiles for industry may be:

- Use in staff training
- Nutrient profiles may support the food industry by improving recipes and promoting product design and innovations
- Support for argumentation aids in marketing and advertising
- Creating equal conditions of competition.

However, until such a concept finally works, a lot of problems must be solved. Generally, the application of nutrient profiling may change not only with the different categories of foods but also with different age and sex of the population, a fact which has to be studied in more detail in the near future. Moreover, in special cases, exceptions will be necessary, e.g. in dietary fats and oils and other special items (pickled olives, spices, anchovy, etc.).

References

1 Draft European Parliament Legislative Resolution on the Council Common Position for Adopting a Regulation of the European Parliament and of the Council on Nutrition and Health Claims Made on Foods (9858/3/2005 – C6-0018/2006 – 2003/0165 (COD).

2 Deutsche Gesellschaft für Ernährung eV (German Nutrition Society, DGE) (ed): Vollwertig Essen und Trinken nach den 10 Regeln der DGE. www.dge.de, Art Nr 122402© 2004.

3 USDA: Nutrition and your health: dietary guidelines for Americans. www.nal.usda.gov/fnic/dga/dguide95.html

4 Deutsche Gesellschaft für Ernährung eV (German Nutrition Society) (ed): Ernährungsbericht 2004. Bonn, DGE Medien Service, 2004.

5 FDA Code of Federal Regulations, Title 21 – Food and Drugs (vol 2), Chapter I – FDA, Department of Health and Human Services, Part 101 – Food Labeling – Table of Contents, Subpart A – General Provisions, Section 101.14: Health Claims: General Requirements and Subpart E – Specific Requirements for Health Claims. Washington, US Government Printing Office, revised as of April 1, 2002.

6 Daily Reference Values (DRVs), based on 2,000 calories a day for adults and children over 4 only. http://www.fda.gov/fdac/special/foodlabel/drvtabl.html

7 Heseker H, Heseker B: Nährstoffe in Lebensmitteln. Frankfurt/Main, Umschau Zeitschriften Verlag, 1999.

8 Rayner M, Scarborough P, Williams C: The origin of Guideline Daily Amount and the Food Standard Agency's Guidance on what counts as 'a lot' and 'a little'. Public Health Nutr 2003;7:549–556.

9 Hammink J, Van den Berg H, Breedveld B: Criteria for the nutritional evaluation of foods: the Netherlands tripartite classification model; Netherlands Nutrition Center, Poster 18th International Nutrition Congress, Durban, Sept 2005.

10 Netherlands Nutrition Center: Criteria for the nutritional evaluation of foods, the Netherlands tripartite classification for foods. www.voedingscentrum.nl, 2005.

11 Rayner M, Scarborough P, Boxer A, Stockley L: Nutrient profiles: development of final model. Final Report. British Heart Foundation Health Promotion Research Group, Department of Public Health, University of Oxford, Dec 2005. http://www.food.gov.uk/healthiereating/nutlab/nutprofmod

12 FSA: Nutrient Profiling, Agenda Item 7, http://www.food.gov.uk/multimedia/pdfs/fsa051004.pdf, Oct 13 2005.

13 Rayner M, Scarborough P, Stockley L: Nutrient profiles: options for definitions for use in relation to food promotion and children's diets. Final Report. British Heart Foundation Health Promotion Research Group, Department of Public Health, University of Oxford, Oct 2004. www.food.gov.uk/multimedia/pdfs/nutrientprofilingfullreport.pdf

14 Erbersdobler HF: 2005 Nährwertprofile – Ansichten und Aussichten. Ernähr Umsch 2005;52:348–353.

15 Nutritional Characterisation of Foods: Science based approach to nutrient profiling. Report from the ILSI Europe Workshop, Mallorca, April 25–27, 2006. Eur J Nutr Suppl (submitted).

Annals of
Nutrition & Metabolism

Ann Nutr Metab 2007;51(suppl 2):50–53
DOI: 10.1159/000103568

Published online: July 30, 2007

Impact of Food Processing on the Implementation of Dietary Guidelines

G. Frisch I. Elmadfa

Institute of Nutritional Sciences, University of Vienna, Vienna, Austria

Key Words

Community nutrition · Nutrient content optimization · Professional food processing

Abstract

Community nutrition offers a wide range of possibilities to increase the health-promoting potential of meals. Yet, the potential for increasing the nutritional quality and the health-promoting value of food is not fully utilized, particularly in the setting of canteen kitchens. There are more than 3,000 canteen kitchens in Austria, and their meal service may have a considerable impact on the well-being of approximately 1.5 million consumers. Nutrient-based dietary guidelines act as an interface between theory and practice and form the basis for the development of recipes and for professional menu planning. In this regard, food processing plays a central role for the implementation of dietary guidelines. It is a common challenge for nutritionists and cooks to provide meals that are both tasty and healthy. Optimization with regard to the health value of community nutrition means reorientation and is composed of two steps: (1) optimization of nutrient and energy contents of menus through a well-balanced diet composition and (2) best-possible preservation of nutrients through professional food processing. Nutrient content optimization in food needs to consider all areas of community nutrition, including a target group oriented nutrient calculation, as well as planning and development of tasty and healthy meals and the highest possible nutrient retention during the entire food preparation process.

Introduction

It is estimated that seven out of ten deaths in Austria were caused by nutrition-related diseases in 2004 alone. In order to reduce this number, comprehensive health-promoting programs are necessary (fig. 1) [1].

The Austrian Nutrition Report 2003 showed insufficient intake levels of some nutrients across all age groups despite an abundance of food and an excessive energy supply. Thus, it is necessary to optimize the availability of individual nutrients already on the level of food processing [2].

The potential for increasing the nutritional quality and the health-promoting value of food is not yet utilized. A close collaboration between nutritional sciences, food industry and catering systems appears to be a promising approach for the optimization of meals with regard to their nutritional value.

According to an investigation carried out by Gastro Obsearcher in cooperation with the Association of the Austrian Food Caterers (AGÖ Arbeitsgemeinschaft Grossküchen Österreichs) in 2005, there are more than

KARGER

Fax +41 61 306 12 34
E-Mail karger@karger.ch
www.karger.com

0250–6807/07/0518–0050$23.50/0

Accessible online at:
www.karger.com/anm

Georg Frisch
Institute of Nutritional Sciences, University of Vienna
Althanstrasse 14, AT–1090 Vienna (Austria)
Tel. +43 1 4277 54911, Fax +43 1 4277 9549
E-Mail georg.frisch@univie.ac.at

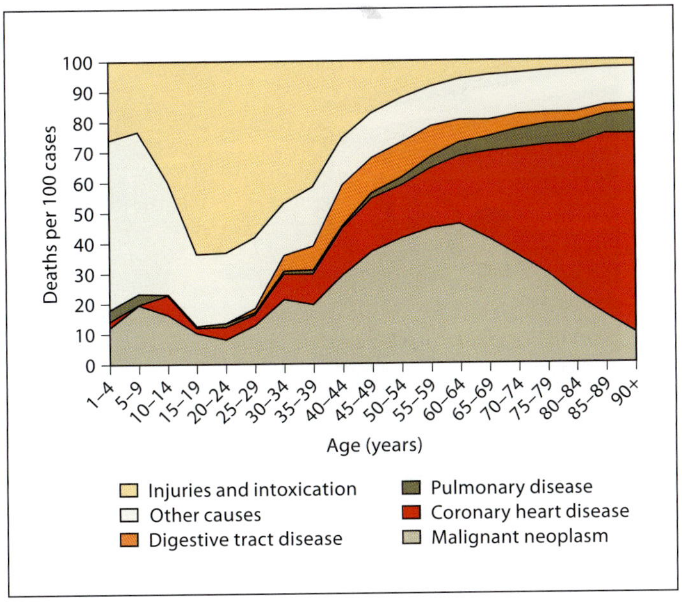

Fig. 1. Causes of deaths in Austria, 2004 [1].

3,000 canteen kitchens in Austria [3]. The work done in these kitchens has a considerable impact on the well-being of approximately 1.5 million eaters. Three quarters of the people utilizing the services of community nutrition regularly are employees (1.2 million). A remarkable share – at least from a health-promoting dimension – is also shown in children (380,000) and the elderly (70,000) [4].

Nowadays, it is becoming more and more common for people to use the services of community nutrition for a long time span – from kindergarten to retirement home. Thus, it is evident that community nutrition plays an important role in public health.

Optimization of the Nutritional Health Value Requires Reorientation

During the past years, community nutrition and catering systems have been rapidly technologically modernized; nevertheless, the nutritional quality of the produced meals has not improved to the same extent. A possible explanation could be that many food manufacturers and cooks still think that meals need to be tasteful but not expensive. Thus, the nutritional value of food plays only a minor role.

Optimization with regard to the health value of community nutrition means reorientation. The sensoric properties and the health value of the food should be of equal importance. This is a common challenge for nutritionists and cooks.

The nutrient-based intake recommendations represent the interface between scientific research and cooking practice. The reference values for energy and nutrient intake, for example the DACH reference values [5], provide important and practical guidelines for the planning of menus, the improvement of formulas, and the development of new recipes. They form the basis for the implementation of marketing strategies, such as the distribution of nutritional information among staff members and consumers. Furthermore, nutrient guidelines represent an essential part of food data bases which are the basis for nutrient analyses. The German Food Code and Nutrient Data Base (BLS = Bundeslebensmittelschlüssel Version II.3) – including raw ingredients, semi-finished food, convenience products and ready-to-eat meals – is linked up to the ingredients of recipes in almost all enterprise resource planning softwares.

As adequate amounts of nutrients available for the organism imply adequate amounts of nutrients contained in foods, it is of utmost importance to guarantee the best possible preservation of nutrients throughout all the steps of food processing. Thus, food processing plays a central role for the implementation of dietary guidelines.

The Two Steps of Nutrient Optimization

Nutrient optimization is composed of two steps, namely (1) optimization of nutrient and energy content of menus through a well-balanced diet composition, and (2) the best possible preservation of nutrients through professional food processing.

The first step is to revisit and evaluate existing meal plans based on the reference values of the target group. Subsequently, data are scrutinized for nutrient-optimizing possibilities with regard to the target group. After analyzing the improved menu plan, the new concept is to be implemented into practice (fig. 2) [6].

Nutrient content optimization in phase 1:

- Rearrangement of menu plans
  - Change of menus as a whole
  - Change of menu components
- Improvement of existing recipes
  - Change of certain ingredients
  - Change in food composition
  - Change of portion size
- Development of new recipes.

For the improvement and the development of new recipes it is necessary to consider that menus with an opti-

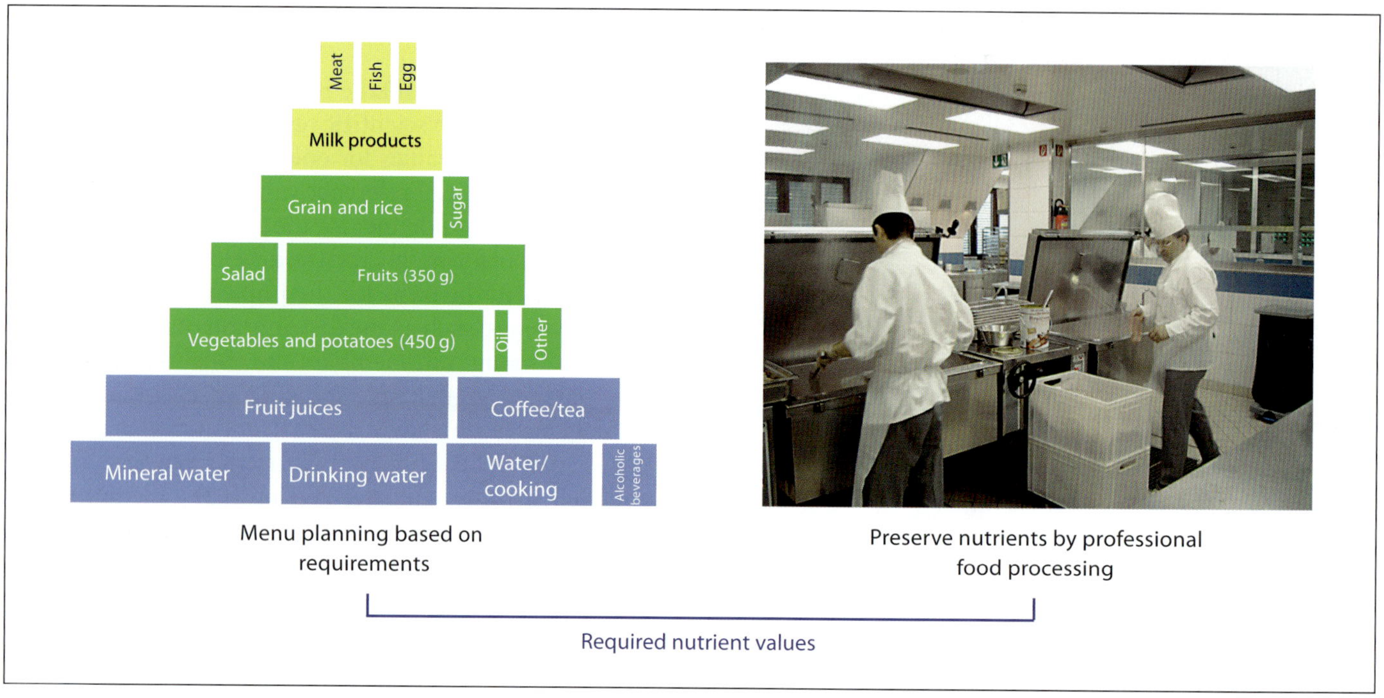

Fig. 2. Nutrient optimization concept [6].

mized health-promoting value still have to be tasty, otherwise these menus will not be chosen by the consumers. For the planning of menus it is advisable to avoid that these 'healthy meals' have to compete with favorite dishes that are poor in nutrients and high in energy.

Nutrient degradation occurs at all steps of food processing [7]. The identification and minimization of this nutrient loss during food processing – including all steps from purchase of raw materials to consumption of the prepared foods – represents phase II of nutrient optimization.

Promising approaches are the selection of ingredients, the storage, preparation, and cooking technologies, the organization of working processes, and the time for keeping the food warm. Skilled staff members play thereby an essential role.

Utilization of nutrient-preserving technologies:

- Protective packaging for stocking of intermediate products
- Cooking equipment with a big capacity for shortening the cook time
- Cooking equipment with automatic time and temperature monitoring and regulation
- Quick chilling systems for hot intermediate products
- Quick chilling systems for ready meals.

The potential for nutrient content optimization can be found at all steps of food processing: already during purchasing it is possible to consider the nutrient density of the raw ingredients. The ratio of fresh food to convenience food, as well as the level of processing of convenience food, has a considerable impact on the nutrient content of the prepared meal. Maintaining the cold chain is essential when taking over the goods. Ingredients delivered in damaged packages and goods nearing their expiry date usually show lower nutrient contents than fresh food. Short storage time of ingredients and optimal storage conditions (adequate temperature and humidity, light and oxygen shield) additionally improve the preservation of nutrients. Furthermore, packages should be sealed carefully again after opening.

The time between consecutive food preparation steps (e.g. peeling, cutting, etc.) should be kept as short as possible in order to protect intermediate products from external influences. Due to the large and unprotected surface area these semi-finished foods are increasingly subject to nutrient degradation caused by oxygen, light, and heat. Adequate cutting utilities prevent the additional squeezing of cells which can protect tissues against enzymatic degradation.

Cooking equipment with automatic temperature monitoring and modern cooking technologies allow for a nutrient-sparing and fat-saving preparation of meals. Through the application of shock-cooling systems it is possible to avoid lengthy and nutrient-consuming storage at relatively high temperatures.

A nutrient optimization concept facilitates the practical application of the guidelines considering:

- Personnel to be made familiar with the applied concept
- A balanced relationship between health value and sensoric properties of the meals
- Nutrient optimization by improving menu planning and recipes
- High nutrient retention during food preparation
- Nutrition information for consumers.

Objective: to produce meals which are tasty *and* healthy.

The following resources and actions are necessary for the implementation of a nutrient optimization concept:

- Cooperation between nutritionists and cooks
- Education and information of the personnel involved (e.g. training of staff members)
- Use of food data bases for the justification of the choice
  - Development of an adapted recipe book
- Adaptation of resources (time, staff, raw material) to the menu plan and the developed recipes
- Employment of nutrient-sparing technologies
- Implementation of quality management systems.

The sensoric properties and the health value of meals are equally considered at all steps of decision-making.

The DGE (German Nutrition Society) and ÖGE (Austrian Nutrition Society) Award a Seal of Quality to Canteen Kitchens That Succeed in Improving the Nutritional Quality of Their Menus

External evaluation and accreditation may act as a strong motivational signal for improving the health value of meals in community nutrition. This is especially true when nutrient optimization is required to gain a competitive edge in tendering procedures. Furthermore, the compliance with these regulations can be used as an internal management instrument. External quality requirements are often associated with public awards which strongly contribute to internal and external image building.

Major evaluation criteria for obtaining the DGE and ÖGE Seal of Quality are: (1) compliance with the DACH reference values for nutrient intake; (2) compliance with recipes throughout food processing, and (3) minimizing nutrient loss throughout meal preparation.

To summarize, community nutrition offers a wide range of possibilities to increase the health-promoting potential of meals. Nutrient-based intake recommendations act as the interface between theory and practice and form the basis for the development of recipes and professional menu planning. Thus, nutrient optimization requires a close cooperation between nutritionists and cooks. Nutrient calculation and food processing are vital for an adequate nutrient content of meals. Nutrient optimization in food needs to consider all areas of community nutrition including a target-group-oriented nutrient calculation, as well as planning and development of tasty and healthy meals and the highest possible nutrient retention during the entire food preparation process. External evaluation and accreditation play an important motivational role for nutrient optimization in community nutrition.

References

1 Statistik Austria, Statistisches Jahrbuch 2006. http://www.statistik.at/fachbereich_03/totenbeschau/allgemein.shtml

2 Elmadfa I, Freisling H: Österreichischer Ernährungsbericht 2003. Institut für Ernährungswissenschaften der Universität Wien, Bundesministerium für Gesundheit und Frauen, 2003.

3 Obergantschnig H: Küchenmanager als anerkannter Beruf. Studie Gemeinschaftsverpflegung, Gastro Obsearcher, Hotel und GV Praxis. Heft 1, Jan/Feb 2006.

4 Nohel CH, Payer H, Rützler H: 2. Lebensmittelbericht Österreich. Die Entwicklung des Lebensmittelsektors von 1995 bis 2002. Bundesministerium für Land und Forstwirtschaft, Juli 2003.

5 DGE, ÖGE, SGE, SVE: Die DACH-Referenzwerte für die Nährstoffzufuhr. Frankfurt am Main, Umschau Braus Verlagsgesellschaft, 2000.

6 Elmadfa I, Frisch G: Die Umsetzung der DACH-Referenzwerte für die Nährstoffzufuhr in der Gemeinschaftsverpflegung. Diplomarbeit am Institut für Ernährungswissenschaften, Universitätsbibliothek Wien, 2002.

7 AID: Nährstoffveränderungen bei der Lebensmittelzubereitung. Bonn 2004.

Author Index Vol. 51, Suppl. 2, 2007

Annals of Nutrition & Metabolism

Subject Index Vol. 51, Suppl. 2, 2007

KARGER

Fax +41 61 306 12 34
E-Mail karger@karger.ch
www.karger.com

Accessible online at:
www.karger.com/anm

Guidelines for Authors

Annals of
Nutrition & Metabolism

Submission

The journal consists of the following sections:

1. Editorials
2. Review Articles
3. Original Papers
4. Case Reports
5. Announcements

Manuscripts written in English are considered and should be submitted online at

www.karger.com/anm

Should you experience problems with your submission, please contact

Prof. Dr. I. Elmadfa
Institute of Nutritional Sciences
University of Vienna
Althanstrasse 14
A–1090 Vienna (Austria)
Tel. +43 1 4277 54901
Fax +43 1 4277 9549
E-Mail ernaehrungswissenschaften@univie.ac.at

Conditions

All manuscripts are subject to editorial review. Manuscripts are received with the explicit understanding that they are not under simultaneous consideration by any other publication. Submission of an article for publication implies transfer of the copyright from the author to the publisher upon acceptance. Accepted papers become the permanent property of 'Annals of Nutrition and Metabolism' and may not be reproduced by any means, in whole or in part, without the written consent of the publisher. It is the author's responsibility to obtain permission to reproduce illustrations, tables, etc. from other publications.

Arrangement

Title page: The first page of each paper should indicate the title, the authors' names, the institute where the work was conducted, and a short title for use as running head. NB: Authors wishing to preserve the phonetic meaning of diacritics (PubMed reduces diacritics to their root characters) must spell their names accordingly when submitting manuscripts (e.g. Müller should be Mueller).

Full address: The exact postal address of the corresponding author complete with postal code must be given at the bottom of the title page. Please also supply phone and fax numbers, as well as e-mail address.

Key words: For indexing purposes, a list of 3–10 key words in English is essential.

Abstract: Each paper needs an abstract of up to 200 words. It should be structured as follows:

Background/Aims:	What is the major problem that prompted the study?
Methods:	How was the study performed?
Results:	Most important findings?
Conclusion:	Most important conclusion?

Footnotes: Avoid footnotes.

Tables and illustrations: Tables are part of the text. Place them at the end of the text file. Illustration data must be stored as separate files. Do not integrate figures into the text. Electronically submitted b/w halftone and color illustrations must have a final resolution of 300 dpi after scaling, line drawings one of 800–1,200 dpi.

Color illustrations
Online edition: Color illustrations are reproduced free of charge. In the print version, the illustrations are reproduced in black and white. Please avoid referring to the colors in the text and figure legends.
Print edition: Up to 6 color illustrations per page can be integrated within the text at CHF/USD 760.00 per page.

References: In the text, identify references by Arabic numerals [in square brackets]. Material submitted for publication but not yet accepted should be noted as 'unpublished data' and not be included in the reference list. The list of references should include only those publications which are cited in the text. Do not alphabetize; number references in the order in which they are first mentioned in the text. The surnames of the authors followed by initials should be given. There should be no punctuation other than a comma to separate the authors. Preferably, please cite all authors. Abbreviate journal names according to the Index Medicus system. (Also see International Committee of Medical Journal Editors: Uniform requirements for manuscripts submitted to biomedical journals. N Engl J Med 1997;336:309–315. www.icmje.org)

Digital Object Identifier (DOI)
S. Karger Publishers supports DOIs as unique identifiers for articles. A DOI number will be printed on the title page of each article. DOIs can be useful in the future for identifying and citing articles published online without volume or issue information. More information can be found at www.doi.org.

Examples
(a) Papers published in periodicals: Chatel J-M, Bernard H, Orson FM: Isolation and characterization of two complete Ara h 2 isoforms cDNA. Int Arch Allergy Immunol 2003;131:14–18.
(b) Papers published only with DOI numbers: Theoharides TC, Boucher W, Spear K: Serum interleukin-6 reflects disease severity and osteoporosis in mastocytosis patients. Int Arch Allergy Immunol DOI: 10.1159/000063858.
(c) Monographs: Matthews DE, Farewell VT: Using and Understanding Medical Statistics, ed 3, revised. Basel, Karger, 1996.
(d) Edited books: DuBois RN: Cyclooxygenase-2 and colorectal cancer; in Dannenberg AJ, Dubois RN (eds): COX-2. Prog Exp Tum Res. Basel, Karger, 2003, vol 37, pp 124–137.

Author's Choice™

With this option the author can choose to make his article freely available online against a one-time fee of USD 2'500.00. This fee is independent of any standard charges for supplementary pages, color images etc. which may apply. More information can be found at www.karger.com/authors_choice.

Page Charges

There are no page charges for papers of 3 or fewer printed pages (including tables, illustrations and references). Each additional complete or partial page is charged to the author at CHF/USD 290.00. The allotted size of a paper is equal to approx. 10 manuscript pages (including tables, illustrations and references).

Proofs

Unless indicated otherwise, proofs are sent to the first-named author and should be returned with the least possible delay. Alterations made in proofs, other than the correction of printer's errors, are charged to the author.

Reprints

Order forms and a price list are sent with the proofs. Orders submitted after the issue is printed are subject to considerably higher prices.

KARGER

Fax +41 61 306 12 34
E-Mail karger@karger.ch
www.karger.com

The Guidelines for Authors are available at:
www.karger.com/anm_Guidelines

Declarations

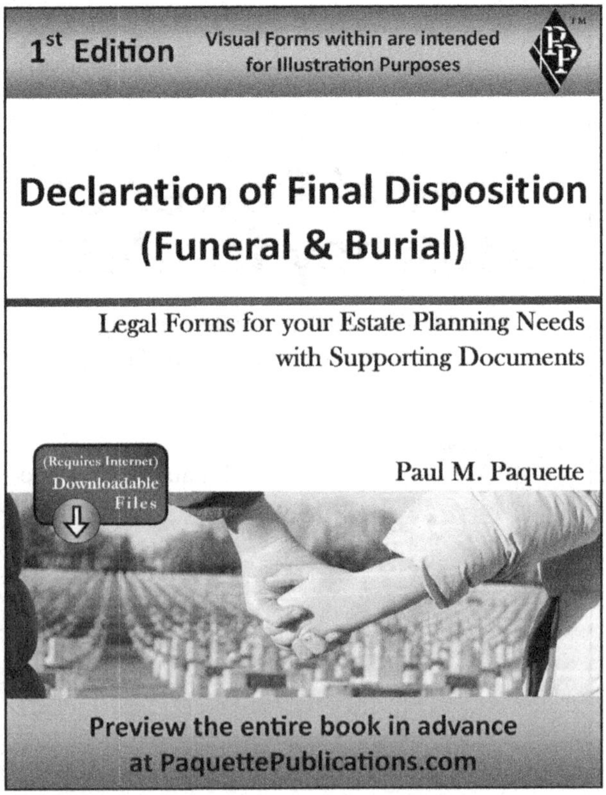

Declarations

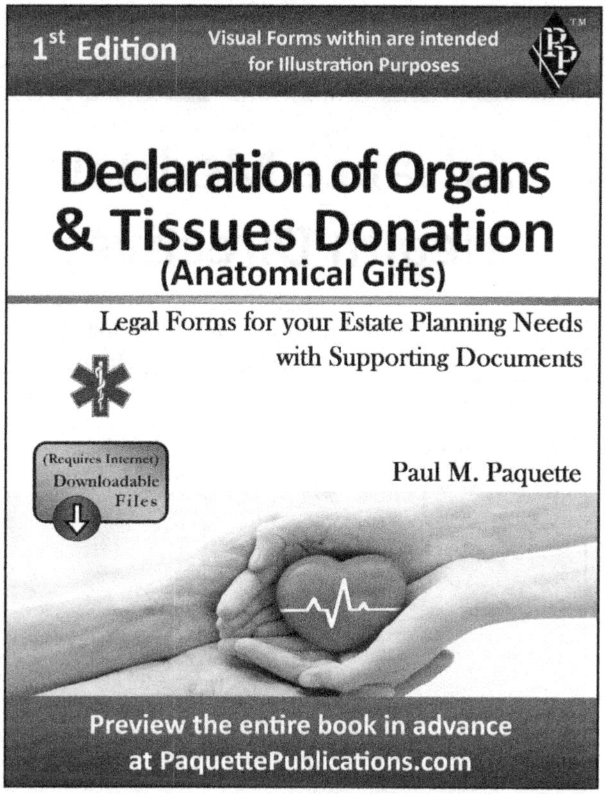

Relationship Agreements

Relationship Agreements

Relationship Agreements

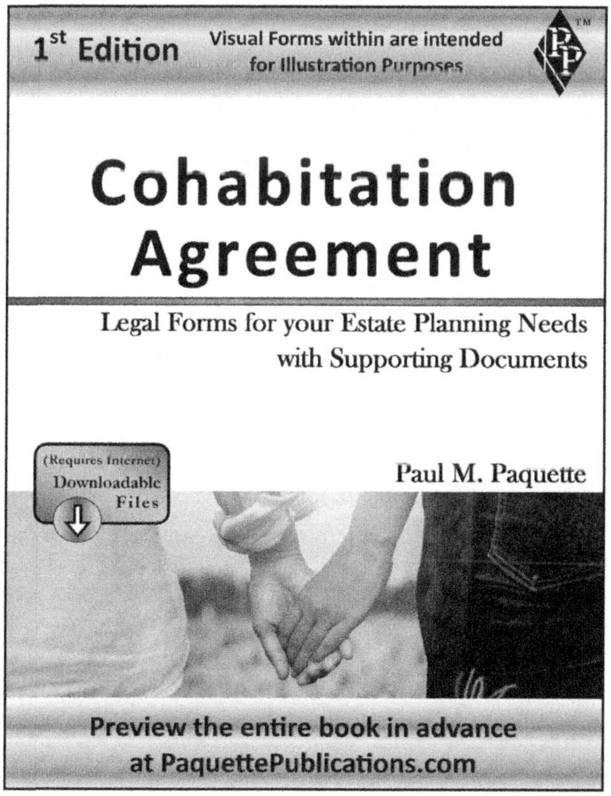

Relationship Agreements

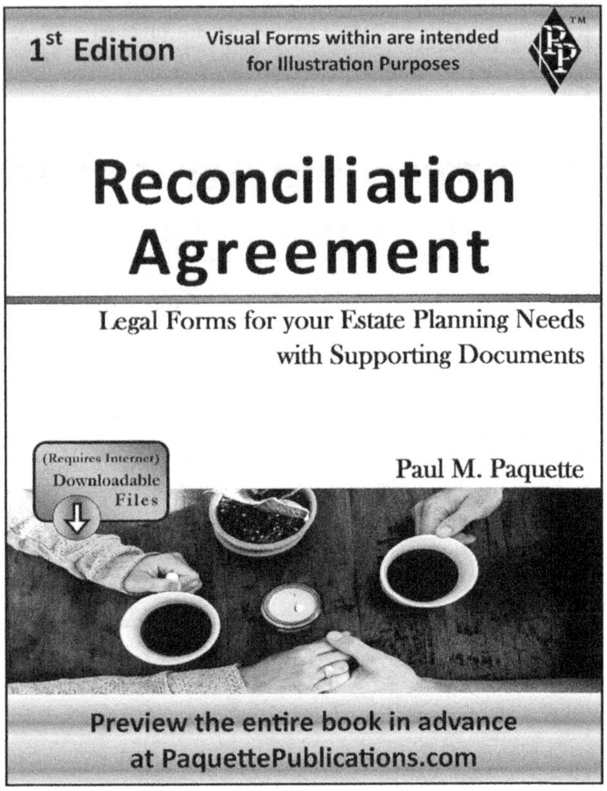

Word Processor Program

The Purchaser or User can open, edit, print, and save the Digital Files using a word processing program. Popular Word Processing Programs include Microsoft Word, WordPerfect, Open Office, and Libre Office. All word processing forms come in the following (**DOC**, **DOCX**, **ODT**). Paquette Publications does not offer the Word Processing Program or Technical Support for the Word Processor Program.

Portable Document Format (PDF) Editor

The Purchaser or User can open, edit, print, and save the Digital Files using a PDF Editor program. Popular PDF Editor include Adobe Acrobat Pro, Nitro Pro 11, PDF-Xchange Editor, Master PDF Editor, and CutePDF Writer. Paquette Publications does not offer the PDF Editor or Technical Support for the PDF Editor. Adobe Acrobat Reader is free to download at **www.adobe.com**. **Please Note:** due to encryption standards of **256-bit AES**, this PDF is only backward compatible (read/write) with **Adobe Acrobat X or later**.

Printer Setting

Depending on the printer's capabilities and settings, the Digital File (**DOC**, **DOCX**, **ODT**) may need editing to prevent cut-off text from occurring near the margins. The easiest way to solve this problem is to verify that the printer and margin settings are correct. If you still have issues, edit/fill out the Digital File, convert it, and save it as a **PDF** file. PDF files are easier, more versatile, and present fewer errors when printing. Paquette Publications does not offer Printing capabilities or Technical Support for Printers.

Modification Recommendation

If the Purchaser or User plans to make changes to the Digital File in question, it is highly advantageous that the Purchaser or User utilize the (**DOC**, **DOCX**, or **ODT**) file format. If the Purchaser or User is content with the choice selection, provisions, and options currently available with minor changes (if applicable), then it is highly recommended that the Purchaser or User utilize the **PDF** file format.

Downloadable Digital Files

Legal Disclaimer:

The Forms presented in this Book are available as a direct download from **www.paquettepublications.com**. In adherence with **Copyright Laws**, **Licensing Agreements**, and **Legal Disclaimers & Waivers**, these forms and their files are for personal and immediate family use only.

Legal Disclaimer required by the Distributor:

This Book does not have digital files (**PDF**, **DOCX**, **DOC**, and **ODT**) attached and does not contain fillable files or forms. As a courtesy, the Author has provided an Active URL Link to download the digital file (**PDF**, **DOCX**, **DOC**, **ODT**); however, the Purchaser or User will need to have an internet service connection, computer access, appropriate software, printing capabilities, and the ability and willingness to access the Author's website.

File Format Options:

The Digital Files are in the following formats: Microsoft Office Word (**DOC**, **DOCX**), Adobe Acrobat (**PDF**), and Open Office (**ODT**). The **PDF**, **DOCX**, **DOC**, and **ODT** are fillable and editable with the appropriate computer software.

Consent Checkbox:

To download these Digital Files (**PDF**, **DOCX**, **DOC**, and **ODT**), the Purchaser or User must first consent to the following Terms and Conditions (**Legal Disclaimer & Waivers**, **Licenses & Trademark**, and **All Rights Reserved**) which are present on the website.

Terms & Conditions

☐ **By checking this box, the Purchaser or User has read and agree to the following:** Legal Disclaimers & Waivers, License & Trademark, **and** All Rights Reserved **before downloading these file(s) (PDF, DOCX, DOC, ODT).**

submit

Download Links:

Using an internet browser (**Chrome**, **Edge**, **Explorer**, **Firefox**, and **Safari**), the Purchaser or User can download the Forms (**PDF**, **DOCX**, **DOC**, and **ODT**) presented within this Book from the following active URL Links:

Power of Attorney:
- https://paquettepublications.weebly.com/tc---cgpoa21---poa12.html

Supporting Documents:
- https://paquettepublications.weebly.com/tc---cd---support.html

Worksheets:
- https://paquettepublications.weebly.com/agrave-la-carte.html

Forms (Miscellaneous):
- https://paquettepublications.weebly.com/tc---all---forms---misc.html

Forms (Recommended):
- https://paquettepublications.weebly.com/tc---all---forms---recommended.html

Appendices & Glossary:
- https://paquettepublications.weebly.com/free---appendix.html

Extras:
- https://paquettepublications.weebly.com/free---extra.html

CD / DVD

This CD / DVD is designed (built and formatted) to work with Microsoft's Windows Operating System but should also work on Apple's Macintosh Operating System. **Please Note:** Paquette Publications does not provide technical support for computers or operating systems.

Print on Demand (POD) book
will have no CD / DVD available.

Please refer to the "**Downloadable Digital Files**" page to access the file(s) (PDF, DOCX, DOC, and ODT).

Please Return Disc with Book

Made in the USA
Monee, IL
05 November 2026

67845295R10118